AF805566

Urban Sustainability

Editor-in-Chief

Ali Cheshmehzangi, School of Architecture, Design and Planning, The University of Queensland, Brisbane, QLD, Australia

The Urban Sustainability Book Series is a valuable resource for sustainability and urban-related education and research. It offers an inter-disciplinary platform covering all four areas of practice, policy, education, research, and their nexus. The publications in this series are related to critical areas of sustainability, urban studies, planning, and urban geography.

This book series aims to put together cutting-edge research findings linked to the overarching field of urban sustainability. The scope and nature of the topic are broad and interdisciplinary and bring together various associated disciplines from sustainable development, environmental sciences, urbanism, etc. With many advanced research findings in the field, there is a need to put together various discussions and contributions on specific sustainability fields, covering a good range of topics on sustainable development, sustainable urbanism, and urban sustainability. Despite the broad range of issues, we note the importance of practical and policy-oriented directions, extending the literature and directions and pathways towards achieving urban sustainability.

The series will appeal to urbanists, geographers, planners, engineers, architects, governmental authorities, policymakers, researchers of all levels, and to all of those interested in a wide-ranging overview of urban sustainability and its associated fields. The series includes monographs and edited volumes, covering a range of topics under the urban sustainability topic, which can also be used for teaching materials.

Ali Cheshmehzangi

AI Agency in Climate Emergency

Springer

Ali Cheshmehzangi
School of Architecture, Design
and Planning
The University of Queensland
Brisbane, QLD, Australia

ISSN 2731-6483 ISSN 2731-6491 (electronic)
Urban Sustainability
ISBN 978-981-95-4191-1 ISBN 978-981-95-4192-8 (eBook)
https://doi.org/10.1007/978-981-95-4192-8

© The Editor(s) (if applicable) and The Author(s), under exclusive license to Springer Nature Singapore Pte Ltd. 2025

This work is subject to copyright. All rights are solely and exclusively licensed by the Publisher, whether the whole or part of the material is concerned, specifically the rights of translation, reprinting, reuse of illustrations, recitation, broadcasting, reproduction on microfilms or in any other physical way, and transmission or information storage and retrieval, electronic adaptation, computer software, or by similar or dissimilar methodology now known or hereafter developed.
The use of general descriptive names, registered names, trademarks, service marks, etc. in this publication does not imply, even in the absence of a specific statement, that such names are exempt from the relevant protective laws and regulations and therefore free for general use.
The publisher, the authors and the editors are safe to assume that the advice and information in this book are believed to be true and accurate at the date of publication. Neither the publisher nor the authors or the editors give a warranty, expressed or implied, with respect to the material contained herein or for any errors or omissions that may have been made. The publisher remains neutral with regard to jurisdictional claims in published maps and institutional affiliations.

This Springer imprint is published by the registered company Springer Nature Singapore Pte Ltd.
The registered company address is: 152 Beach Road, #21-01/04 Gateway East, Singapore 189721, Singapore

If disposing of this product, please recycle the paper.

AI-generated image, developed by Ali Cheshmehzangi, in Plainland, QLD, 29 July 2025

For those whose absence became a source of resilience—
those who perished in the shadows of every war and conflict of 2025,
a year that bore witness to more than 60 burning frontlines,
each a wound in the conscience of our world.
To the unnamed, the unheard, the erased—
but to our complicit silence, our selective outrage, our inaction dressed as neutrality.
whose lives were lost not only to weapons,
This book bears witness to what many chose not to see.
The inhumanity we normalized,
the suffering we scrolled past, the justice we deferred, and the empathy we rationed.
Let this dedication be a crack in that silence—
a remembrance, a reckoning,
and a refusal to forget what must never be accepted.
It shall be written not just in the margins of history,
but in the deep, dark ink of humanity's failure to protect its own.

Preface

> *"AI is not a silver bullet, but there are exciting opportunities to apply AI to help accelerate the transition to net zero and support climate resilience. The opportunity AI offers on climate change is currently under-appreciated by governments".*
>
> *— Peter Clutton-Brock, Co-founder, Centre for AI & Climate*

We are part of an evolution, coconsciously, sub-coconsciously, and unconsciously. From the age of computers, setting up those groundbreaking machines and new virtual brains to the world of internet, we have been part of a process of welcoming and accepting the AI at a large scale. With the earlier establishment and evolution of mind and the extended stored information and data—in the form of an endless e-library—that we have co-created or contributed to, AI seems to be just the next inevitable step. But the question is what next? What could we do right or wrong in what may be ahead of us?

Let us confront a reality we can no longer soften with euphemism: this is no longer "climate change", it is rather a 'climate emergency'. The world is on fire, and we are still debating where to place the thermostat. This is not a time for gentle words. This is not a time for patient observation, polite optimism, or techno-utopian fantasy. This is a time to confront the unbearable truths of our era: the planet is warming faster than we are acting. The systems we once trusted to govern, to plan, to adapt, are all unravelling. Climate instability is no longer a future projection; it is a lived reality. Yet our response remains disturbingly slow, fragmented, and hollow.

We have created a type of cognitive infrastructure that, with careful governance, could help us see what we refuse to see and act where we continue to delay. Meanwhile, a new force is subtly emerging, (hopefully) not to replace us, but to reveal us. This is artificial intelligence, not the AI of sensationalist headlines or dystopian imagination, but a more subtle and embedded intelligence: silently parsing satellite images of deforestation, optimizing energy grids in milliseconds, tracking emissions from power plants that governments fail to report, and listening to digital cries for assistance in the wake of floods, droughts, and fires. We must overcome the weaponization of AI, which has been so quickly incorporated into contemporary military equipment, and recover its potential for something much more pressing and significant: addressing the dire consequences of our climate emergency. We are lacking in insight, not weaponry. Let AI be used to detect, defend, and provide a habitable future with accuracy, speed, and moral clarity—not to destroy.

This book is not a celebration of technology. It is not an ode to AI. It is a demand. A demand that we stop treating this climate emergency as a purely ecological crisis or a political inconvenience. It is a systemic rupture. And AI, in all its evolving forms, must now be treated not merely as a tool of convenience or control, but as an agent of response.

We must be clear: AI won't save us. There is no neural network for justice, no algorithm for bravery. However, there is something particularly potent about AI's ability to both magnify our collective capabilities and act as a mirror to our failing institutions. Today's climate action requires scale. It requires velocity. It requires vision, collaboration, feedback, and flexibility across systems that are now too fragmented, sluggish, and unyielding to change. AI, a novel type of actor, a sort of creative agency, that becomes essential in this situation.

Indeed, AI agency and applications in climate change research are underappreciated, underestimated, and dangerously underused.

In the chapters that follow, I present a transdisciplinary journey through the emerging landscape of *AI agency in climate emergency*. Before a single drop of water falls, you will learn about digital twins of cities that mimic flood conditions. While businesses continue to deny it, you will read about machine learning algorithms that identify methane leaks from orbit. You will learn about AI systems that expose the greenwashing concealed in ESG disclosures and platforms that elevate public voices in climate assemblies. But more importantly, you will encounter a deeper question that underpins this entire work: *What kind of intelligence do we need for the planet we have?*

This question haunts me. Because heat, sea level rise, and biodiversity loss are not the only aspects of the dilemma. It concerns our inability as a society to control complexity, an inability to consider multiple systems, to hear over time and to take action on a large basis. It's a cognitive crisis, to put it briefly. Furthermore, AI needs to be a component of a new planetary intelligence architecture that is intended for discernment rather than dominance if it is to have any significance in this setting.

This book advocates for a fundamental reinterpretation of AI as a scaffold rather than a rescuer or so-called saviour as we wrongly perceive. It advocates for a framework for improved perception, improved coordination, and improved inclusivity. Let us use our creativity a bit more. The concept of "smartness" must give way to a relational, flexible, open, and inclusive understanding of intelligence. Intelligence that works, learns, and listens to promote regeneration rather than extraction. This vision is not technological. It is a moral and political position.

Let me be blunt: the status quo is fatal. We do not need another conference to tell us that emissions are still rising. How many of those rambling discussions and nonsense keynotes have we listened to in many money-generating—but yet often very useless—conferences around the world? We do not need another glossy strategy document that treats equity as a footnote. What is the use all those generic documents that are created with little—it not zero—impact on the realities of our society? What we need is a fundamental shift in how we structure response capacity. This is the true promise of AI, not automation for convenience, but augmentation for survival. Not control over nature, but co-adaptation with it.

Some may read this and ask: why place so much faith in machines, when what we need is human resolve? My answer is simple: I place no faith in machines. I place faith in our ability to shape them, innovate them, co-create them, and utilise them in the best possible way. To design systems that extend—not erode—human agency. To build infrastructures of intelligence that reflect not just what we know, but who we choose to be.

Others may ask: why risk creating another layer of techno-colonialism or surveillance? My answer is again clear: if we do not govern AI with justice at its core, that is exactly what we will get. That is why this book insists on AI that is transparent, inclusive, and if I may against my own belief, democratically governed. AI that is embedded in communities, aligned with Indigenous knowledge, and responsive to frontline realities.

We are already seeing glimpses of this future. AI that supports regenerative agriculture by analyzing soil carbon levels in the Cerrado. AI that flags discrepancies in corporate carbon accounting, exposing fraudulent offsets. AI that helps design equitable adaptation finance strategies for cities on the brink. These are not imaginary futures, they are happening now. But they remain scattered, underfunded, and often politically marginalized.

That is what this book seeks to change. It is both a call to arms and a map of possibility. A way of seeing AI not as a product, but as a process. Not as a fixed system, but as a field of interaction between code and climate, governance and justice, data and desire. This is where our work lies: not in resisting the future, but in co-creating it with innovation, courage, and better understanding of what it actually entails.

Today, we are at the intersection of several crises, but we are also at the verge of something even more spectacular. Let the climate crisis inspire us to create new institutions if it has shown the shortcomings of our current ones. In this view, intelligence is something humans need to regain, not merely something AI provides.

Because whether AI will be employed in the fight against climate change is not the main question. It is already. Whether it will be used in a timely manner, with integrity, and for the futures that really count is the question. The futures that encompass everyone, not just those who own the servers or wrote the programming, but also those who must deal with the fallout.

This book is for them. For the thinkers, designers, activists, planners, scientists, policy-makers, and everyday citizens who refuse to surrender to fatalism. Those who still believe we can design our way toward something better. Who understand that the real intelligence we need is not artificial, it is collective, and is made of us by us.

May this book provoke, inspire, and disturb.

May the book, with all its provoking thoughts and ideas, challenge your assumptions. May it equip you to ask bolder questions. And may it remind you, again and again, that climate action is not only a technical challenge. It is a choice. A choice about the systems we build, the values we embed, and the worlds we dare to imagine.

Let us begin, before it is too late.

August 2025

Prof. Ali Cheshmehzangi
Head of School of Architecture, Design and Planning
The University of Queensland
Brisbane, QLD, Australia

Acknowledgements I acknowledge that several images and diagrams featured in this book have been significantly enhanced through the integration of AI-assisted design tools. The process involved numerous iterations, adjustments, and refinements (including many unusual—but seemingly essential—one-to-one arguments with the tools) to ensure each visual not only aligned with the thematic and conceptual aims of the work but also conveyed complex ideas with clarity and precision. These AI-generated contributions, developed after many trials, have helped bridge creative intent with technical accuracy, producing visuals that complement and elevate the narrative of the book. I found the whole process fascinating yet quite tiring from time to time, but it worked very well in generating images that are not just relevant, but informative and compelling. The content and concepts represented in these images, however, are entirely my own and were not generated by AI in any form. Rather, the AI tools served solely as a means of visual production, providing valuable assistance in translating my ideas into clear and engaging visuals that support and enhance the material presented here.

More importantly, I would like to sincerely acknowledge the quiet toll my unusual rhythms of writing often take on those around me. The late nights, the early mornings, the moments of stillness interrupted by sudden thoughts; these are the conditions under which my books take form every single time. And I acknowledge it happens more often than I plan every time/year/month/week. I remember Juan's wise words, spoken with equal parts humor and truth: "*It looks like writing books has become a daily drug for you*". This is probably a humorous but right statement. I cannot disagree with that, but yet keep a smiling face thinking about it more and more. Indeed, writing has become more than a habit for me. It has become a compulsion of the thinking mind, a healing stirred at odd hours, in fleeting moments while walking, dreaming, or simply being in unfamiliar and inspiring places.

This acknowledgment, then, is not conventional. It is a small, sincere gesture to honour both the drive that fuels this work and the patience of those who witness its more chaotic and jagged edges. Yes, great effort goes into shaping each book, and into contributing meaningfully to the fields I deeply care about. But there is also a cost, often borne quietly by those closest to me. To Juan, and others who understand this strange devotion to words and ideas, thank you. You see not only the outcome, but the obsession and each journey of writing a new book; and most important of all, you endure it with grace. This truth deserves to be said, and I dare to record this utterly sincerely here.

Competing Interests The author has no competing interests to declare that are relevant to the content of this manuscript.

About This Book

AI Agency in Climate Emergency is a daring and topical examination of how artificial intelligence is becoming a pressing—and yet integrated—force in the battle against climate change rather than a far-off solution. This book repositions AI as a new type of actor inside our ecological, political, and social systems, rather than just as a computational tool, in an era characterized by mounting crises, such as rising oceans, failing infrastructure, disappearing forests, and growing inequality.

The book covers the entire spectrum of climate response, from energy decarbonization to foresight and early warning systems, from precision agriculture and ecosystem monitoring to green finance, carbon markets, civic engagement, and behavioural change. It is organized into six thematic sections and twenty integrated chapters. AI's developing agency—its ability to recognize patterns too big for human awareness, to adjust in real time, and to close the crucial gap between knowledge and action—is what ties these disparate fields together.

However, the subject matter of this book is not technological determinism. It is a book on imagination, values or ethics, and governance. It makes the case that justice, transparency, and solidarity—rather than just speed—must be the guiding principles of AI's future in climate action. It promotes AI that is inclusive, adaptive, accountable, and purpose-aligned with the welfare of the planet—AI that learns from context as well as data.

The book is intended for academics, professionals, decision-makers, and citizens who recognize that climate change is a crisis of systems rather than merely emissions. Furthermore, both artificial and human intelligence must advance to a level of wisdom in the impending urgent shift. This book provides a convincing road map for achieving that goal.

Contents

About the Author

Ali Cheshmehzangi is a Professor and Head of the School of Architecture, Design and Planning (ADP) at The University of Queensland. He is the Brisbane Olympics Committee member for Games Legacy and Growth Sub-Committee as well as the Expert member of the 2032 Games led by UQ. He has been in the World's top 2 percent field leader since 2021, recognised by Stanford University. He is among the top 20 global scholars in the urban sustainability research area.

Ali has a wealth of experience, expertise, and a strong commitment to innovation and excellence in the fields of architecture, design, and planning, the three core disciplines of his School at UQ. With a career spanning over two decades, he has made significant contributions to the academic and professional communities, with a focus on sustainable and environmentally conscious design. Prior to joining UQ, Ali held several strategic leadership and senior managerial roles, such as Vice-President for International Engagement and Global Partnership, Founding Director and Head of the Center for Innovation in Education and Research, Head of the Department of Architecture and Built Environment, Founding Director of the Urban Innovation Lab, Director of a university-wide Teaching and Learning platform, Director of International Research Network for Rural and Urban Development, Founder and Head of Research Group for Sustainable Built Environment, co-director of university-wide research priority areas, Director of Center for Sustainable Energy Technologies, and Interim Director of Digital Design Lab. So far, Ali has published over 500 journal papers, articles, conference papers, book chapters, and reports. This is his 50th academic book, alongside his many poetry collections that he has written in Parsi/Persian and English. Some of his previous academic books have received awards at the national, provincial, and municipal levels. He also has received international awards and recognition for his research on urban resilience studies and sustainability research, as well as a 2018 Vice-Chancellor's award for his impactful contribution to higher education.

Chapter 1
Awakening the Machine Mind: AI as a Force for Planetary Response in Dealing with Our Collective Climate Emergency

1.1 A Climate Emergency, a Cognitive Gap

Our era is characterized by thresholds, including generational reckonings, social disruptions, and planetary tipping points. The likelihood of a climate emergency is no longer remote. It is present, compounding, and unfolding unevenly across geographies and lives. We are seeing the breakdown of the stability that formerly structured the rhythms of civilization, from unheard-of wildfires and tumbling floods to increasing heat stress and food hunger. However, despite the increasing sophistication of research, the gap between knowledge and action continues to be persistently large. This is a crisis of cognition as well as one of politics and resources. We are becoming less adept at understanding, processing, and reacting to the complexity and speed of climate change.

The mismatch between the emergency's scope and the decision-making mechanisms entrusted with handling it is the cognitive gap of the climate period. Our environment was slower and more predictable when our institutions, models, and mental models were formed. They are designed to optimize for linear issues, plan in short cycles, and act after the event. All these presumptions are contradicted by climate change. It is very unpredictable, linked, and non-linear. As we argue, it speeds up. As we pause, it adjusts. Not only is our infrastructure at risk, but so is our creativity.

The result is paralysis masquerading as 'advancement'; promises on emissions that do not match the actual levels of emissions, policies that fall short of meeting the needs for adaptability, sprawling cities that boast resilience, and financial flows that disregard reality in favour of green signals. Additionally, communities are forced to carry the most load with the fewest resources, particularly those on the fringes. We're not stupid. The capacity to mobilize it at the proper scale, speed, and depth is what we lack. We don't have systems that can detect patterns beneath noise, adjust in real time, or sense before failure.

© The Author(s), under exclusive license to Springer Nature Singapore Pte Ltd. 2025

A. Cheshmehzangi, *AI Agency in Climate Emergency*, Urban Sustainability,
https://doi.org/10.1007/978-981-95-4192-8_1

Enter artificial intelligence (AI), not as a silver bullet (Holmström et al., 2024), but as a new form of cognition that, if carefully aligned, can help bridge the gap between awareness and action, or can assist close the awareness-action gap. AI offers more than automation or prediction. AI is more than just prediction or automation. It is not a panacea. At its best, it can comprehend complexity, identify signals in uncertainty, and model outcomes before they materialize, making it a system-level amplifier of human intellect. Importantly, it can model what might be, allowing foresight and adaptation in a world where precedent is no longer a trustworthy guide.

The promise of AI in the context of climate change, however, is not to replace human agency but to extend it—helping communities, cities, and institutions think across scales, across silos, and across time. This is AI as a climate co-agent: integrated into urban systems, embedded in public infrastructure, aligned with justice-centered goals, and accountable to those it serves. Instead of AI as a technological saviour, augmented intelligence is intended to be an adaptive strategy.

However, the future is not assured. AI has the potential to both close and exacerbate the cognitive divide. It can centralize power, reinforce inequality, reify skewed datasets, and divert attention with flimsy accuracy. It can optimize systems that ought to be abolished, serve the powerful, and monitor the weak. In summary, AI has the same potential drawbacks as any human technology. Its climate utility will be decided not by its capabilities alone, but by the values and institutions that shape how—and for whom—it is deployed.

The conflict between crises and cognition is where this book starts for that reason. We must comprehend the landscape that AI is entering before we can investigate how it can map wildfires, optimize energy grids, model climate-responsive cities, or track deforestation from space. This is a tale of ecological ethics, societal adaptability, and planetary governance, not merely invention. Better tools are not enough to address the climate crisis. It necessitates the adoption of new, distributed, anticipatory, and care-based ways of thinking, interacting, and doing.

Lastly, it is crucial to note that AI will not solve climate change. However, if we view the cognitive gap as an opportunity rather than a weakness, it can become a vital partner in the change we need to make. An encouragement to think differently, act more quickly, and create systems that are wise as well as intelligent.

1.1.1 Beyond Human Limits: Complexity, Speed, and Scale in the Age of Disruption

This section diagnoses the climate emergency as a crisis of delayed cognition, fragmented governance, and systems that are too sluggish to respond, in addition to being an environmental crisis. It presents the idea of the cognitive gap, which is the growing gap between the rate of climate change and the ability of conventional institutions, instruments, and human intuition to adapt. The sub-section examines how the magnitude of feedback loops, cascading system failures, and interrelated risks create this gap, laying the groundwork for the requirement for augmented intelligence and systems-level understanding.

The climate crisis is a rapidly developing disturbance of human settlements, planetary processes, and governmental institutions rather than only a slow-moving environmental hazard. It goes beyond what is typically understood and controlled in all areas—economic, social, and ecological. Its feedback-driven, non-linear dynamics are frequently undetectable until they cross thresholds. Ocean currents change when ice sheets melt. Rainfall patterns are altered by deforestation thousands of kilometers away. Mortality risks are increased by urban heat islands. Disruptions to the supply chain have an impact on continents. These occurrences are not unique. They are indicators of stressed systems that are interrelated, accelerating, and poorly understood.

The primary problem with the climate emergency is not merely that it is occurring, but that it is occurring more quickly and erratically than our institutions and instruments can manage. In a twenty-first-century risk environment, we are using infrastructure from the nineteenth century and governance from the twentieth century. The intricacy and unpredictability of climate-driven events frequently overwhelm the cognitive capacity of even the most advanced enterprises. Reactions that are delayed are common, not unusual. Physical reality frequently takes precedence over policy. Political inertia clashes with scientific consensus. Furthermore, linear solutions soon prove insufficient in situations where climate impacts cascade; droughts escalating migration, heatwaves taxing electricity grids, and floods causing disease outbreaks.

The cognitive gap is essentially a structural disconnect between our existing decision-making systems' ability to understand and respond to the reality of rapidly occurring changes. Conventional methods depend on post-hoc interventions, compartmentalized expertise, and recurring evaluations. When dangers were slower and more geographically limited, these might have been adequate. However, the speed at which crises are mounting in the climate period necessitates ongoing sensing, quick learning, and ongoing re-calibration. We require tools that can not only predict the future but also change in real time as it develops.

The boundaries are human, not just technological. Our ability to recognize new dangers, foresee ripple effects, and deal with uncertainty is hampered by cognitive overload, restricted rationality, and institutional inertia. By affecting food and energy, water and conflict, health and shelter, climate change exacerbates these limitations. No one dataset, discipline, or agency can fully understand the situation. This fragmentation is an inevitable feature of global systems with dispersed effects, not a weakness. However, it makes fragmented responses inadequate. More separate solutions are not necessary. Systems that can react to disruption at the speed and scale at which it occurs must be coordinated, flexible, and enhanced with intelligence.

The feedback loops of climate change are particularly insidious. Loss of Arctic ice lowers reflectance, which raises warmth and causes additional ice to melt. Droughts increase the danger of drought by killing plants, which lowers carbon uptake and increases atmospheric concentration. These loops are hard to break once they start. Even worse, their beginning is frequently postponed, making early intervention crucial but challenging to politically defend in the absence of an obvious crisis.

This also applies to social tipping points. The social fabric itself may disintegrate during climate shocks when infrastructure malfunctions or institutional trust declines, exacerbating the physical harm.

These are governance failures resulting from epistemic lag, not merely engineering or environmental issues. Institutions lose their ability to adapt when they are unable to recognize stress signals in a timely manner or when they do not take appropriate action. In this situation, resilience is more than just getting back up after a setback. It involves early detection, efficient coordination, and decisive action. And to do that, systems that can learn more quickly than the crises they are meant to solve are needed, not just human intuition.

This is where AI comes into play, not as a rescuer but as an essential addition to our finite capacity for thought. The ability to process large datasets in real time, spot subtle correlations, model future paths, and make dynamic strategy adjustments are all made possible by AI. Therefore, it is evidenced that AI offers a way to close the cognitive gap—not by replacing human judgment, but by enhancing it with speed, pattern recognition, and foresight—whether it is used to model urban flooding under various climate scenarios, forecast the spread of wildfires using wind and fuel load data, or optimize energy systems under peak demand stress.

AI is particularly effective in settings where uncertainty and complexity are compounded by climate change, where no single person or organization can fully understand the situation. In this case, machine learning models can incorporate several inputs, including social, economic, hydrological, and meteorological data. As circumstances change, they are able to continuously update projections. Instead of static blueprints, they can offer adaptive planning. Importantly, they can also act as a feedback bridge between reaction and reality, converting signals into insight and action from insight.

However, AI is insufficient on its own. Its incorporation into transparent, accountable, and justice-focused systems is what counts. In the absence of this, AI runs the risk of developing into yet another tool of elite control, favouring the wrong outcomes for the wrong individuals. The cognitive divide has an institutional and ethical answer in addition to a technical one. It entails rethinking governance as a distributed, data-driven, and collaborative process. Catching up to the rate of disruption is not the only objective. By creating intelligence systems that can learn across domains, respond before failure, and change with the climate, we can get ahead of it.

In short, the climate emergency is putting more than just our businesses and infrastructure to the test. It is putting our knowledge, reasoning, and reaction to the test. And in this disruptive age, those methods themselves need to be upended—by intelligence that observes more, picks things up more quickly, and serves a large number of people.

1.2 The Rise of Machine Agency: A New Kind of Actor in Climate Systems

A new type of actor is emerging as the climate catastrophe worsens; it is not a human or an organization, but rather a dispersed intelligence that is incorporated into the same mechanisms that we use to comprehend, control, and respond to the disturbances that are taking place on Earth. Once restricted to the back end of scientific computing or corporate optimization, AI is now firmly making its way into front-line roles (Kellogg et al., 2022). It can now be used to detect illegal deforestation from satellite streams, balance electrical grids in real time, predict fires before they start, and even influence the creation or implementation of policies. This change signifies the rise of a new type of operational agency inside climate systems in addition to a technological advancement.

AI taking the place of human decision-makers or attaining some fanciful kind of conscious autonomy is not the goal of this agency. Instead, it is about adaptive algorithms, sensor networks, and machine learning models taking an active role in the logic of climate response. These systems are becoming more and more capable of interpreting complicated inputs, coordinating dynamic reactions, and influencing decisions that directly or indirectly affect the health of the planet. They obfuscate the distinction between actor and tool in the process. They wait. They control. They acquire knowledge. They take action.

An interactive, embedded notion of machine agency is replacing the conventional view of AI as a passive data processor. For instance, machine learning methods are currently used in wildfire control to anticipate ignition risk zones days ahead of time by combining topography, fuel maps, satellite data, and weather forecasts. Public warnings, evacuation plans, and the pre-deployment of firefighting resources are all influenced by these forecasts. In this case, AI is actively organizing resource flows, changing institutional preparedness, and impacting life-saving outcomes in addition to offering insight. Its agency comes from influencing other people's behaviour rather than from acting independently.

AI functions inside the real-time choreography of demand, distribution, and generation in energy systems in a similar manner. Based on weather models, it forecasts solar and wind output, modifies power dispatch in response to fluctuating needs, and instructs electric vehicles or smart appliances to draw or release electricity at the best times. These choices are made continuously, independently, and at a rate that is much faster than human ability. In this situation, AI serves as the brains behind decarbonized energy networks, balancing affordability, emissions, and resilience in the face of volatility and uncertainty.

AI-powered remote sensing is revolutionizing the way we identify and address environmental degradation in land systems (Cheshmehzangi, 2026). Large amounts of satellite data are processed by machine vision algorithms to detect changes in land cover, illicit logging, and even the first indications of ecosystem stress or desertification. By strengthening local stewardship and upending illicit supply chains, these tools enable law enforcement and civil society to take action with never-before-seen speed and accuracy. AI systems can even provide regions risk rankings based on probable future degradation in certain situations, allowing for proactive conservation measures as opposed to reactive damage control.

The way AI is currently incorporated into action logic is what unites all of these examples. It's not only observational anymore. It is both operational and anticipatory. Its outputs are real-time signals, alerts, and directives that influence resources, tactics, and behaviour rather than being remote reports. Even though its presence is not always apparent, AI is successfully co-governing alongside humans in several climate-relevant domains, such as infrastructure and agriculture.

This expanding machine agency has significant ramifications for our understanding of responsibility, accountability, and power in relation to climate action. Who makes the choice when a neural network improves crop irrigation in an area that is prone to drought? Whose preferences win out when the energy system of a smart city blocks power draw in order to maintain grid stability? Whose preconceptions are embedded in the algorithm when an AI model is used to score climate risk for finance or insurance? The values, interests, and institutions of the cultures that create and use machine agency are reflected in it; it does not live in a vacuum.

Furthermore, the distribution of machine agency is not uniform. High-quality data, technical know-how, and sophisticated computational tools are available to some areas, organizations, and communities (Pourroostaei Ardakani & Cheshmehzangi, 2023). Others don't. Because of this asymmetry, unless AI's agency is planned, tracked, and regulated with purpose, its expanding involvement in climate governance runs the risk of escalating already-existing disparities. We must now examine computer actors functioning within vital systems with the same level of scrutiny that we would a human bureaucrat working without supervision or redress.

To be clear, there is nothing fundamentally wrong with AI's emergence as a climate actor. In reality, it is a potent ally in managing the climate emergency because to its ability to evaluate complexity, model futures, and adjust in real time (Brevini, 2023). However, we need to stop thinking about AI as a tool and start thinking of it as infrastructure, a component of the fundamental framework that senses, models, coordinates, and implements decisions. Understanding machine agency entails understanding the politics woven into code, the data-driven ethics, and the algorithmic design-driven results.

This chapter presents the main idea of the book, which is that AI is now influencing climate action rather than merely aiding it. By doing this, it has evolved into a new type of actor in our ecological destiny. One that has to be revealed. One that needs to be handled carefully. And one whose expanding agency must be based not only on engineering excellence but also on planetary boundaries, climate justice, and democratic ideals.

We will examine the various ways AI is utilizing this new agency in a variety of systems, including energy, cities, agriculture, forests, finance, and government, in the upcoming chapters. However, acknowledging that we are no longer alone in the architecture of climate response is where it starts. Our thoughts are shared with machines. The resilience—or fragility—of our planet's future will be determined by the form of that collective thought.

1.3 What Kind of Intelligence Do We Need? Framing the AI We Want for the Planet We Have

If the climate emergency has revealed anything, it is that intelligence of any kind is not neutral. The attitudes, presumptions, and orientations that underlie intelligence—whether it be artificial or human—influence how we perceive and handle problems. The stakes are significantly higher in the domain of AI. The question is no more whether we will utilize AI to combat climate change, but rather what sort of intelligence we need to maintain a habitable planet as AI systems progressively influence our perceptions of risk, resource allocation, and solution-imagining.

This chapter makes the case for a significant reinterpretation. AI shouldn't be viewed as a general-purpose optimizer or computing resource. It needs to be created as a type of purpose-aligned intelligence that takes into account the intricate social, ecological, and moral demands of our day (see Fig. 1.1). There is more to the climate catastrophe than just a technical issue with emissions and measurements. Because social, ecological, and economic systems are collapsing, it necessitates intelligences that can handle diversity, interconnection, and uncertainty. The core objectives of climate justice, intergenerational equity, and ecological regeneration must be served by the AI we require, not merely abstract efficiency.

First among the required orientations is *relational intelligence.* Climate change is a systems problem, cascading across scales and sectors. It cannot be solved using reductionist metrics or comprehended in a vacuum. Therefore, AI systems need to be built with the ability to perceive relationships—to place data in the context of lived surroundings, cultural customs, and ecological dynamics. This entails modelling interdependencies rather than just limited optimization, such as how changing transportation policy alters air quality, how changing water use impacts energy systems, or how forest destruction concurrently depletes local livelihoods and global carbon

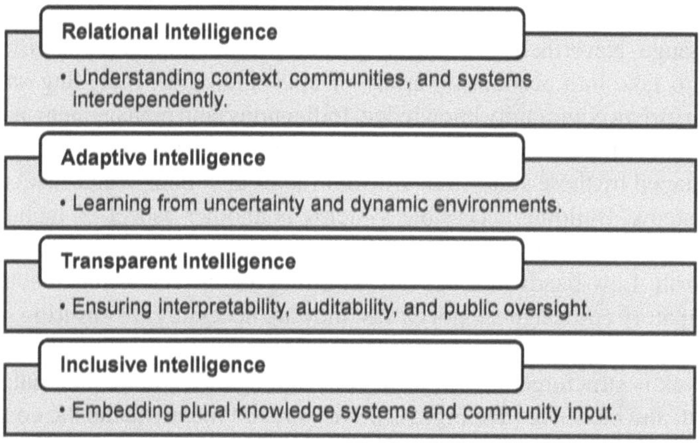

Fig. 1.1 Four types of intelligence brought into discussion here for the narrative of AI agency in climate emergency

stocks (Rawat et al., 2024). Relational intelligence makes it possible for AI to identify patterns as signals woven across the world's interwoven web of infrastructure and life, rather than as abstract data points. According to this perspective, AI must develop into a systems thinker who is highly contextual, cross-sectoral, and perceptive of both local and global patterns.

Second is *adaptive intelligence.* The climatic age is characterized by volatility in terms of resources, governance capability, weather, and politics. A world of non-linear change will not be able to support static models and rigid assumptions. AI systems need to be constantly learning, revising their models in response to changing circumstances and fresh information. This calls for institutional flexibility as well as technical flexibility, such as the ability to retrain neural networks or modify feedback mechanisms, so that AI systems can operate in dynamic coalitions, hybrid governance structures, and unpredictable policy contexts. To put it simply, AI needs to go from predictive rigidity to responsive fluidity. The objective is to move with agility and humility through the unpredictable future, not to control it.

Third, and foundational to trust, is *transparent intelligence.* AI shouldn't be a mystery with unintelligible forecasts and cryptic algorithms. Explainability and auditability are essential in climate-relevant sectors because actions impact ecosystems, economics, and lives. Stakeholders need to be able to comprehend the methods used to arrive at findings, weigh data, and manage ambiguity. Model design, data provenance, system governance, and human oversight are some of the layers at which transparency needs to be planned. Furthermore, openness guarantees that mistakes can be identified, prejudices can be addressed, and choices can be linked back to their morally and rationally sound foundations. It is not just about visibility. In high-stakes situations like emissions reporting, deforestation alerts, or adaptation finance, where errors in judgment might result in resource misallocations or a decline in public confidence, this is especially crucial.

Finally, we must insist on *inclusive intelligence.* Historically, the prevailing AI models have given preference to specific knowledge systems that are data-driven, technocratic, and frequently disconnected from the populations most impacted by climate change. Nevertheless, solutions created without human input will not work. AI needs to take into account a variety of epistemologies, including smallholder knowledge, urban community knowledge, Indigenous land management, and climate youth activism. Model assumptions, design priorities, and decision-making processes must be shaped by these sources of wisdom rather than being "add-ons" to technological systems. Building accessible systems is another aspect of inclusion, as it applies to frontline communities worldwide as well as affluent institutions in the Global North. Low-bandwidth options, multilingual interfaces, open-source tools, and participatory co-design are not extras; they are necessary for ensuring legitimacy and justice in AI-led climate governance.

This book is structured around these four intelligences, i.e., relational, adaptive, transparent, and inclusive (See Fig. 1.2). They are not merely idealistic concepts, but rather workable ideas that are already being adopted in a variety of climate-related fields. We are seeing a new generation of climate intelligence systems that represent more than just computational power, from digital twins that simulate urban cooling

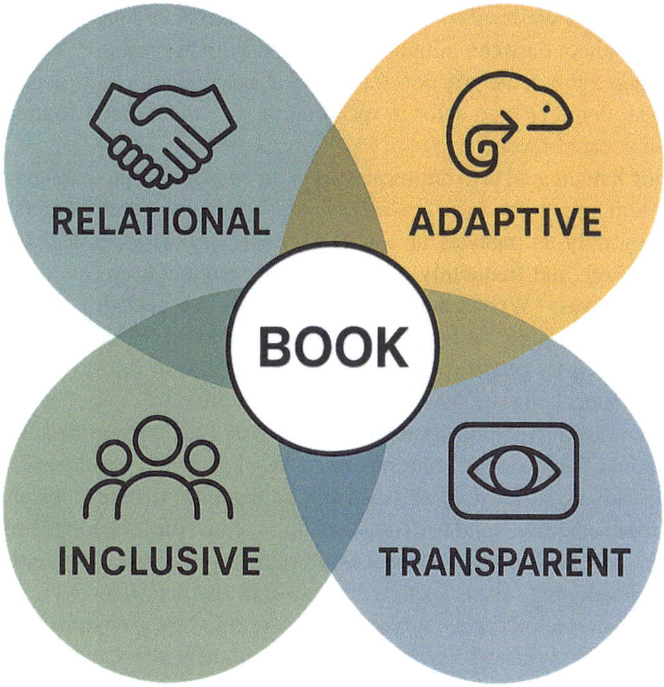

Fig. 1.2 Four intelligences that shape the structure of this book

strategies to algorithmic sentiment analysis of climate protests, from AI-assisted reforestation projects in the Congo Basin to predictive irrigation in drought-prone India. They provide a sense of purpose.

Our aim is not to romanticize AI, nor to assume that it will "solve" the climate crisis. Instead, it aims to show how the needs of a warming planet are already influencing many types of AI, whether consciously or unconsciously. We contend that how, by whom, and for what purposes AI will be utilized in climate action is more important than whether it will be employed at all. The futures we enable will be shaped by the type of intelligence we create today. The true frontier, as we approach a global transition, lies not only in data or code, but also in integrating machine intelligence with the ideals and aspirations of a fair and sustainable Earth.

1.4 The Aims and Architecture of This Book

This book is the result of a straightforward but pressing realization: the climate emergency is a crisis of coordination, time, and thought, not only of emissions, ecosystems, or infrastructure. We live in a time when the complexity and speed of planetary disturbance are outpacing the ability of our institutions, technologies, and knowledge-making processes to respond. There is a risk and an opportunity in this growing disconnect between climate change and human response. The goal of this book is to shed light on the prospect that AI, if developed with consideration, equity, and foresight, could be a co-actor in speeding up, expanding, and changing climate action at all societal levels.

This book's main goal is to reinterpret AI as an emerging type of agency inside the climate system rather than as an abstract or remote technological force. It is evolving beyond being only an analysis or automation tool to become a decision-maker, a system optimizer, and frequently a co-creator of results. Given this realization, we must ask ourselves, "What kind of intelligence do we need for the world we're entering?" and put aside the hype and anxiety around AI. What ideals, objectives, and design principles ought to direct the incorporation of AI into urban systems, agriculture, finance, climate governance, and daily life?

The book is organized into six main sections that trace the outlines of the climate catastrophe and the systemic solutions made feasible by AI to address these questions. From climate forecasting and catastrophe response to renewable energy coordination, urban adaption, land use transformation, emissions tracking, and participatory governance, each section examines a crucial area where AI is either already present or developing. Each section has several chapters that delve deeply into certain use cases, technological advancements, regulatory environments, and moral dilemmas—always tying AI's potential to practical difficulties and limitations.

This architecture is intentional. As energy is related to land, cities impact food systems, emissions are linked to money flows, and individual actions influence communal destinies, it reflects the interconnectedness of the climate system itself. By organizing the book in this manner, we hope to map a web of agency in addition to providing a list of advances. We aim to demonstrate how AI may act as a cross-domain connecting tissue, facilitating a more intelligent and well-coordinated response to the climate crisis.

This book's emphasis on multiplicity is another of its main goals. The chapters will include instances from Latin America, Africa, Asia, and the Pacific, in addition to the Global North, where AI is being tailored to local contexts, enabling Indigenous stewardship, or closing access gaps in smallholder agriculture. AI for climate action, in our opinion, needs to be ethically regulated, locally based, and globally informed.

This book is ultimately about more than just technology. Vision is key. It is about the future we wish to create and how augmented intelligence can help make it a reality; not as a replacement for political will or group action, but as a means of enabling their more intelligent and equitable manifestation.

1.5 Crossing the Sectors, Centering the Citizen

The goal, organization, and main questions of the book are outlined in this section. From urban resilience to carbon markets, from climate foresight to participatory governance, it outlines the cross-sectoral framework and emphasizes the central theme of intelligent climate systems: putting people and equity first. The subsection describes how each chapter advances knowledge of AI's capabilities for ecosystems, institutions, and regular people dealing with climate stress. This book explores AI's role in creating resilient, regenerative futures rather than just its capabilities.

The purpose of this book is to be cross-sectoral. Neither climate change nor our response should respect disciplinary boundaries. Energy systems, cities, agriculture, banking, land use, and public governance are all covered in the book—not to give a cursory overview, but to show how AI is starting to become a common thread across these fields. Its main focus is not on what AI can do, but rather on what AI makes possible—for ecosystems under stress, for institutions grappling with complexity, and—above all—for individuals negotiating the lived realities of a warming planet.

Each chapter looks at how machine learning, neural networks, and algorithmic systems are being incorporated into climate-relevant infrastructures, ranging from decentralized energy grids to predictive weather prediction, from AI-assisted urban planning to climate-resilient farming and carbon market verification. However, AI is not treated as a neutral toolkit in this work. It poses more profound queries, such as: For whose purposes are these systems intended? What values do their architectures encode? Who stands to gain—and who could lose out? By raising these issues, the book positions AI as a site of moral decision-making, cultural interpretation, and civic design rather than as an unavoidable force.

A fundamental belief that underpins our investigation is that the climate crisis is a human issue as much as an environmental or technological one. As a result, AI cannot be assessed based only on performance indicators. It must be evaluated based on how it affects people's lives, rights, and ability to adapt, act, and prosper. Every chapter in this book adheres to the same main theme: placing the citizen at the center of intelligent climate systems, whether it is talking about algorithmic irrigation in farms affected by drought or providing Indigenous land defenders with real-time deforestation alarms.

This entails realizing that AI is influencing daily life rather than only being used in prestigious policy rooms or Silicon Valley firms (Margetts & Dunleavy, 2024). It can be found in satellite models that forecast future flood locations, apps that make recycling more fun, and smart home systems that encourage individuals to use less energy. However, the distribution of these systems is not uniform. Deep disparities still exist in decision-making authority, computer literacy, bandwidth, and data access. For this reason, this book challenges the terms of inclusion of AI rather than merely mapping its capabilities. Can AI-augmented agriculture involve smallholder farmers? Can the training of urban digital twins be influenced by frontline communities? Can carbon tracking become more accessible through open-source platforms?

The book is divided into six sections, each of which focuses on a crucial nexus between AI and climate action, in order to address these issues. The story starts with climate forecasting, where AI enhances our ability to detect, simulate, and react to catastrophic catastrophes. From there, it shifts to the intelligent restructuring of energy systems. After that, it looks at how AI is improving carbon accountability, rethinking the built environment, safeguarding ecological systems, and—above all—promoting institutional and behavioural change via civic design, nudging, and participation. This reflects the type of integrated intelligence, one that transcends silos, listens intently, and adapts constantly, that is required for climate resilience. The book encourages readers to track AI's traces across these environments, not just to observe its current direction but also to influence its future. It seeks to establish a forum for collective reflection among designers, scientists, policymakers, technologists, and community leaders—on the types of futures we are constructing and the intelligence systems we are integrating into them.

In doing so, this book does not present an automatic salvation dream. It presents a more strong and realistic picture of enhanced climate agency. People, not just machines, are empowered by responsive, inclusive, and transparent systems. In this narrative, AI is not the protagonist of the climate crisis. It is a partner. One that can guide us through the intricacy, rapidity, and unpredictability of this moment—but only if we maintain humanity at the core of its design.

1.6 A Future Worth Knowing: Reclaiming the Climate Imagination

This final and most inspiring section reframes climate action as an act of collective imagination; i.e., one that must be grounded in science but lifted by vision. It presents AI as a narrative accelerator that aids mankind in imagining what justice, sustainability, and collaboration might look like in times of crisis, rather than merely as an analytical instrument (Şerbu & Mârza, 2024). It considers how machine learning may also teach us to learn again—from the world, from data, and from one another—and why an intelligent future needs to be characterized by the values we instil in it rather than just by computation. It ends with the question: Will we create systems that react only after it's too late, or ones that can listen, learn, and take the lead before it's too late, in a world of feedback loops and crashing timelines?

For a long time, climate change has been presented as a problem for science, a problem for policy, a strain on the economy, and, more lately, a battlefield for technology. However, it is fundamentally also a crisis of imagination. Our current situation is a failure to conceive alternatives boldly, cooperatively, and early enough. It goes beyond a failure to limit emissions or prepare for disasters. The window for incremental action gets smaller as the climate gets hotter, ecosystems break down, and timeframes get shorter. We require more than planning in such a situation. We require fresh perspectives. novel approaches to listening. novel approaches to knowledge.

When AI is limited in its scope, it may appear unsuitable for this kind of work. It optimizes systems, finds patterns, and parses data. But in the age of climate change, AI has greater promise not just in terms of processing capacity but also in terms of broadening our ethical and cognitive perspectives. In the end, this book challenges readers to view AI as a mirror, a compass, a catalyst, and a narrative accelerator for a planet in crisis, rather than just as a tool for fixing problems.

AI allows us to model futures we cannot physically visit. Before we cross tipping points, it mimics them. It shows us where cities choke and forests breathe, where floods will occur, where policies will fail, and where, with care and daring, hope may flicker. It can help us practice the future of the world and, if applied with humility and justice, change the future of the world before the practice is realized. But it must be worthwhile to know about this future. It must be based on collaborative foresight rather than reactionary fear; on ecological belonging rather than abstraction; and on shared responsibility rather than technological dominance.

Reclaiming the climate imagination as design, rather than escape, is necessary to get there. As foresight, not as fantasy. We need to imagine AI systems that are able to coordinate, include, question, and co-create in addition to calculating. This means that relational capacity—the capacity to act sensibly across time, across difference, and across systems—should be used to define intelligence rather than just speed or scale.

From the land, the data, the wind, and the wisdom of people who have coexisted peacefully with nature for far longer than any algorithm has existed, we must learn anew—possibly even learn to learn again. We must realize that we cannot be saved by specialized knowledge in the face of a polycrisis (Assadourian, 2025). The urgency of this era can only be met by a distributed intellect that is shared by humans, robots, and ecosystems. And so the final provocation: Will we create systems that don't react until collapse is irrefutable in a world where global feedback loops and cascading dangers rule? Or, before it's too late, will we create systems that can listen, learn, and take the lead?

The following chapters are in six parts, each composed of three chapters, that embody the central arc of the book's narrative: the emergence of AI agency in the climate emergency. Each section examines how intelligent systems are becoming active actors rather than only analytical observers, affecting our perceptions, decisions, plans, and behaviours in a world that is changing quickly. These six components collectively reveal a new form of climate intelligence that is not just quicker or more intelligent but also more responsible, inclusive, and sensitive to the interconnected destinies of people and the planet. This is not a dream of technology. It is a moral, ethical, and political argument. And it starts with the straightforward yet profound query: What sort of intellect do we want to lead the world that we are attempting to save?

References

Assadourian, E. (2025). What comes after the polycrisis? *Global Sustainability, 8*, Article e12.

Brevini, B. (2023). Artificial intelligence, artificial solutions: Placing the climate emergency at the center of AI developments. *Technology and Sustainable Development* (pp. 23–33). Routledge.

Cheshmehzangi, A. (2026). *Generative AI-Powered Urban Digital Twins: Pioneering Environmental Solutions for Sustainable Intelligent Cities*. Springer.

Holmström, A., Ahlmark, D. I., Lugnet, J., Andersson, S., & Ericson, Å. (2024). Cybersecurity and the AI Silver Bullet. In *Artificial Intelligence for Security: Enhancing Protection in a Changing World* (pp. 23–33). Springer Nature Switzerland.

Kellogg, K. C., Sendak, M., & Balu, S. (2022). AI on the Front Lines. *MIT Sloan Management Review, 63*(4), 44–50.

Margetts, H., & Dunleavy, P. (2024). The political economy of digital government: How Silicon Valley firms drove conversion to data science and artificial intelligence in public management. *Public Money & Management*, 1–11.

Pourroostaei Ardakani, S., & Cheshmehzangi, A. (2023). The role of big data analytics in urban systems: review and prospect for smart transport and healthcare systems. *Big Data Analytics for Smart Transport and Healthcare Systems*, 1–26.

Rawat, A., Kumar, D., & Khati, B. S. (2024). A review on climate change impacts, models, and its consequences on different sectors: A systematic approach. *Journal of Water and Climate Change, 15*(1), 104–126.

Şerbu, R. S., & Mârza, B. Ş. (2024). From Crisis to Opportunity: Embracing Sustainable Development Goals and Artificial Intelligence in the Transformative Innovations World. In *Ukraine's journey to recovery, reform and post-war reconstruction: A blueprint for security, resilience and development* (pp. 49–61). Springer Nature Switzerland.

Part I
AI for Climate Foresight: Monitoring, Prediction, and Rapid Response

This section examines how artificial intelligence is improving our capacity to see, forecast, and react to environmental concerns, thereby changing climate foresight. AI makes it possible to project temperature increases, sea level changes, and extreme weather events more precisely using high-resolution climate modelling. Machine learning-powered advanced early warning systems analyze massive sensor and satellite data to identify threats like floods and wildfires at previously unheard-of speeds. Finally, AI-driven disaster response systems facilitate real-time coordination, assisting emergency services in identifying populations at danger and refining evacuation plans. This represents a significant advancement in proactive risk management and climate resilience.

Chapter 2
Reimagining the Climate Lens: AI-Powered Modelling for a World in Flux

2.1 The Limits of Traditional Climate Modelling

The scientific community has been simulating the Earth's climate system for decades using complex but computationally demanding technologies. The most prominent of these are the General Circulation Models (GCMs) and the Regional Climate Models (RCMs), which mimic atmospheric, oceanic, and terrestrial phenomena throughout time using physical equations and boundary conditions (Cash et al., 2005; Chokkavarapu & Mandla, 2019). While RCMs provide a higher resolution across smaller geographic areas and are frequently used to downscale GCM outputs for regional studies, GCMs function at global scales and usually describe climatic phenomena using grid cells that range in size from 100 to 250 km.

The foundation of climate research is made up of these models. They support international policy frameworks, direct long-term emissions scenarios, and serve as the foundation for global assessments conducted by organizations like the Intergovernmental Panel on Climate Change (IPCC). Traditional climate models are rigorous and credible, but they have basic flaws that make it difficult for them to adapt to the growing climate threats of the modern world.

Temporal and spatial resolution is one important limitation. Fine-grained phenomena like urban heat islands, localized flash floods, or regional drought patterns frequently fall below the resolution threshold, even though large-scale weather patterns can be represented very effectively. A 150 km grid cell, for instance, is unable to discern between coastal zones, agricultural plains, and dense metropolitan areas—all of which are impacted by climate change (Gopalakrishnan et al., 2019) in very different ways.

Computational intensity is the second significant constraint. High-resolution climate predictions require a significant amount of processing power and run-time (Mizielinski et al., 2014), frequently spanning days or even weeks on supercomputers, especially those required for scenario testing or real-time risk assessments.

© The Author(s), under exclusive license to Springer Nature Singapore Pte Ltd. 2025
A. Cheshmehzangi, *AI Agency in Climate Emergency*, Urban Sustainability,
https://doi.org/10.1007/978-981-95-4192-8_2

This creates a bottleneck in terms of both generating results and promptly updating models in response to fresh data or rapidly changing environmental conditions.

Most importantly, conventional models frequently lack flexibility when dealing with climate phenomena that are non-linear, rapidly changing, and cascading. Models are not allowed to catch up in the actual world. The predictive power of traditional modelling tools is increasingly challenged by events like compound disasters (cyclones followed by infrastructure collapse), feedback loops (warming accelerated by Arctic ice loss), or unexpected extremes (record-breaking heatwaves, abrupt glacier collapses). This is especially true when these tools rely heavily on historical baselines that may no longer be applicable in a world that is changing quickly.

These drawbacks highlight the pressing need for modelling methods that are faster, more adaptive, and have higher resolution. We need tools that can keep up with the planet's pace—technologies that can forecast impending hazards at neighbourhood or watershed scales in addition to simulating large-scale processes—as climate impacts become more urgent, unpredictable, and spatially diversified. AI steps in at this point, providing a supplementary and potentially revolutionary paradigm in climate foresight.

The following part examines how AI is redefining what is feasible, enabling a more flexible reaction to the intricacies of a warming planet, and uncovering hyperlocal insights. It is also increasing the efficiency of climate simulations.

2.2 AI's Leap Forward: Enhancing Resolution and Speed

In climate research, AI is swiftly emerging as a key instrument for overcoming the basic drawbacks of conventional climate modelling, and it is no longer merely a futuristic idea (Sathvara et al., 2024). This change is primarily due to AI's capacity to speed up processing, compress data, and analyze patterns in ways that physical simulation cannot. AI models mimic and even improve the accuracy of climate projections by learning directly from large datasets, including observational, satellite-based, and simulation-generated data, rather than depending exclusively on solving intricate differential equations over grids of atmospheric and oceanic variables.

Machine learning-driven downscaling is among the most potent innovations. Dynamic downscaling in traditional modelling necessitates nesting high-resolution regional models inside coarser global models, a procedure that is nonetheless time-consuming and computationally costly. On the other hand, machine learning methods like super-resolution generative adversarial networks (SRGANs) and convolutional neural networks (CNNs) are able to "learn" the statistical connections between coarse inputs and fine-scale outputs (Murukesh et al., 2024). Then, without requiring the execution of complete physical simulations each time, these models quickly produce high-resolution spatial forecasts that provide insights into the local climate. Climate variables like temperature, precipitation, and wind speed, for instance, can be projected at sizes as small as 1 km, providing vital information for local disaster preparedness or urban resilience design (See Fig. 2.1).

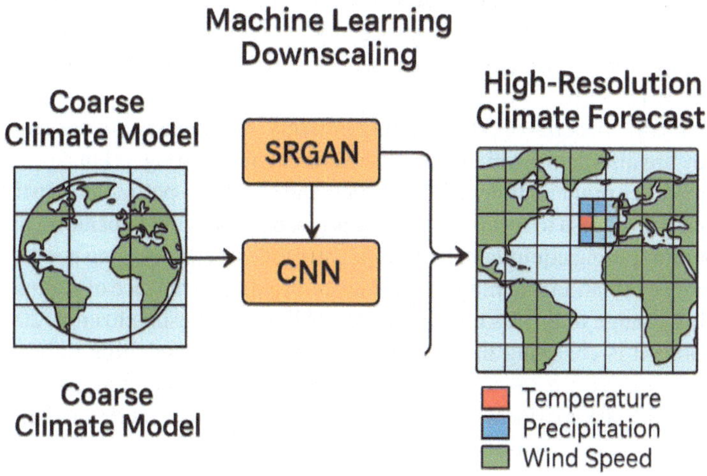

Fig. 2.1 A summary of machine learning-driven downscaling process

Furthermore, deep learning architectures—particularly those designed for spatiotemporal data, such as LSTMs, attention mechanisms, and recurrent neural networks—are especially well-suited for spotting intricate temporal patterns and cascading effects in climate data. This makes it possible to develop models that identify new trends, nonlinearities, and abnormal behaviours rather than just extrapolating historical data—a capability that is becoming increasingly important in our age of climate uncertainty.

At a fraction of the computing cost, FourCastNet, a well-known example created by NVIDIA and partners (Djouama et al., 2024), employs deep learning to generate weather and climate forecasts with an accuracy that is comparable to that of conventional numerical models. FourCastNet uses learnt representations from past simulations and observational data to produce global weather forecasts in seconds rather than hours on supercomputers. This speed opens the door to opportunities for adaptive modelling, near real-time scenario testing, and sector-wide, climate-informed decision-making.

Physics-informed machine learning, in which AI models are trained using both data and the physical principles regulating climate systems, is another expanding application. By bridging the gap between physical realism and data-driven insights (Kong et al., 2023), this hybrid approach enhances generalizability and builds confidence in model results. It also aids in resolving the lack of physical consistency and transparency, which is one of the main objections of pure machine learning models.

In addition to speed and resolution, AI makes it possible to quantify uncertainty and perform ensemble forecasting at a scale that was previously unfeasible. To assess how sensitive projections are to various initial conditions or climate paths, multiple

AI models can be trained and executed concurrently. This helps to educate risk-averse or precautionary strategies—especially in regions sensitive to abrupt climatic shifts—by providing decision-makers with a variety of potential outcomes rather than just one forecast.

Crucially, the benefits of AI modelling in terms of efficiency are not merely technical; they are also democratizing access to climate forecasting. AI makes it possible for low-resource organizations, communities, and regions—particularly in the Global South—to gain access to localized climate projections without depending on centralized supercomputing equipment by significantly lowering processing needs. By facilitating more egalitarian planning and resilience-building in the face of common global issues, this change improves climate justice (Allam & Cheshmehzangi, 2024).

In short, AI is changing the fundamentals of climate modelling rather than just speeding it up. AI creates a new frontier where cities, communities, and policy-makers may act on hyperlocal, near-instant insights about the dangers and futures they confront rather than on out-of-date forecasts or generic patterns by making modelling faster, more scalable, and more spatially aware. The next section focuses on the practical applications of these capabilities, particularly in modelling the compound events, temperature extremes, and sea-level rise that characterize the frontlines of the climate catastrophe.

2.3 Precision Projections: Sea-Level Rise, Temperature, and Extremes

Climate change is not a far-off concept; rather, it is being manifested through abrupt shocks, gradual changes, and escalating extremes. Rising sea levels, rising temperatures, and complex weather events that put stress on human and environmental systems are some of the most urgent aspects. Accurately predicting these phenomena over time, geography, and severity levels is essential for both short-term responses that save lives and safeguard infrastructure as well as long-term planning. In order to provide these projections with the granularity, speed, and adaptability that traditional models frequently lack, AI is suddenly becoming essential.

Take sea-level rise, for instance. The reality on the ground is far more unequal, even though global averages give an idea of trajectory. Land subsidence, ocean currents, glacial isostatic adjustment, and storm surge patterns all have an impact on local sea-level variations. These intricate, multi-source datasets, which include coastal topography, tide gauge records, satellite altimetry, and even societal vulnerability indices, can be fed into AI models to produce maps of flood risk and localized sea level projections. Forecasting the danger of flooding in particular coastal settlements under different climatic scenarios is made possible by deep learning technologies, such as ensemble neural networks, which are especially good at detecting nonlinear correlations between geophysical factors (See Fig. 2.2.).

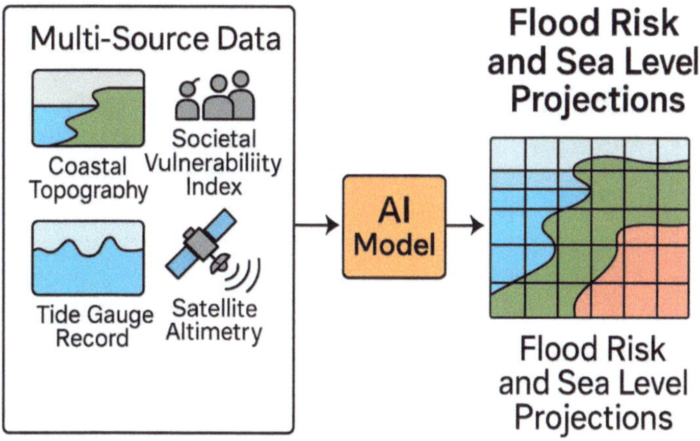

Fig. 2.2 AI and sea level rise matters using multi-source data

AI aids in reorienting attention from general regional averages to neighbourhood-scale thermal reality in relation to temperature rise and heat extremes. Urban micro-climates can differ greatly over short distances due to a variety of factors, including land use, vegetation cover, and building materials. AI-enhanced models can detect heat-vulnerable areas, anticipate urban heat island impacts, and even model the cooling potential of interventions like reflective surfaces or tree planting by using high-resolution satellite imagery, meteorological data, and urban morphology. AI gives city planners evidence-based methods for climate-responsive design by quickly mapping, modelling, and simulating these solutions (See Fig. 2.3).

More critically, AI excels in identifying and forecasting compound and cascading events—those climate shocks that occur in tandem or sequence, amplifying damage and overwhelming systems. These include drought-related heat waves, storms that cause urban flooding, or wildfire seasons that coincide with infrastructure breakdowns. Because these interactions are complicated and unpredictable, traditional models frequently have trouble reproducing them. On the other hand, AI systems that have been trained on real-time inputs and disaster data in the past are able to identify patterns in a variety of dimensions, identify possible overlaps, and model future compound risks with a high degree of temporal sensitivity (Rehan, 2022; Thekdi et al., 2023). AI, for instance, can forecast how heat stress and electricity grid loads are related (Vahedi et al., 2025), or how soil moisture deficiencies may increase the chance of wildfires.

Furthermore, cross-variable learning is also made possible by transformer-based models and attention processes, which are now typical in sophisticated AI systems that link land-use dynamics, hydrological patterns, and weather anomalies. Decision-makers can use probabilistic, AI-informed scenario planning to allocate resources

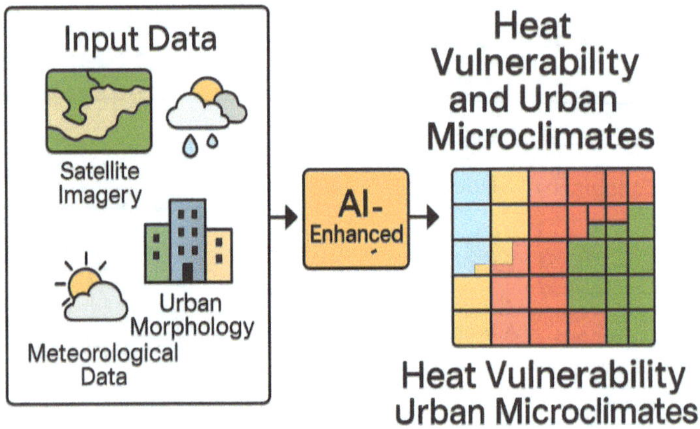

Fig. 2.3 AI maps urban microclimates to target heat vulnerability interventions

for emergency logistics, infrastructure reinforcement, and early warning systems thanks to this multifaceted understanding.AI is also enhancing climate-health forecasting, a growing area of importance. By correlating extreme heat events with health data, energy usage, and air quality indices, AI models are informing real-time alerts for heatstroke risks, respiratory illness upticks, and hospital demand surges. These insights are becoming critical as climate impacts move from the abstract planetary level into daily human consequences.

Ultimately, AI's potential for climate prediction extends beyond technical advancement. It resides in its ability to bring order to chaos by converting massive amounts of data into tangible, useful knowledge. AI gives us the ability to observe earlier, simulate deeper, and act more quickly as heatwaves increase, sea levels rise, and catastrophic occurrences get more intertwined. The third and last section discusses how these insights might be put into practice, specifically how AI-powered climate modelling can be used to inform future-focused and realistically grounded plans (Cheshmehzangi, 2026), policies, and infrastructure choices.

2.4 From Insight to Action: Policy and Planning Implications

Whether driven by machine learning algorithms or simulations based on physics, the real value of climate modelling is found in the decisions it facilitates rather than the models themselves. AI is causing a simultaneous shift in how planners, legislators, and communities approach risk, resilience, and the future as it changes the field of climate foresight (Ojanen, 2025). AI's contribution to climate modelling goes beyond resolution or speed; it also involves strategic clarity. Climate adaptation moves from speculative policy to targeted, time-sensitive action when coastal agencies can see sea-level consequences monthly or when city governments have access to block-by-block heat projections. It is possible to update urban design codes to incorporate accurate thermal profiles. As hazards arise, flood zoning can change dynamically. AI-enhanced probabilistic futures can be used to prioritize infrastructure investments rather than historical averages.

AI-driven climate projections are already being incorporated by lenders and insurers into risk assessments and pricing models in the financial sector, impacting real estate markets, insurance coverage for susceptible regions, and credit allocation. In particular, in areas that face compound challenges from heat, water scarcity, and coastal exposure, governments are starting to match long-term infrastructure investment and land-use strategies with AI-generated climate scenarios.

Most significantly, AI makes climate planning more accessible. The dissemination of high-resolution, data-rich forecasts via civic dashboards or open-access platforms gives local authorities and citizens more control over adaptation plans (Alshahrani, 2025). Previously hindered by technological obstacles, participatory planning becomes more inclusive thanks to accessible risk maps, AI-generated "what-if" scenarios, and visual simulations that turn complexity into observable futures.

AI-enhanced modelling does, however, come with risks and obligations, just like any other potent instrument. To guarantee legitimacy and confidence, model transparency, ethical data sourcing, and algorithmic accountability must be given top priority. Policymakers must also acknowledge the inherent uncertainty and fluidity of climate futures and refrain from placing an undue emphasis on predictive certainty. AI is a compass that continuously adjusts itself in response to new information and patterns; it is not a crystal ball.

In this sense, the integration of AI into climate modelling is more than just a technical advancement. It signals a move toward adaptive governance, in which planning is responsive, tactics change in tandem with the data, and foresight is iterative. AI has the potential to not only predict the future but also to shape it, guiding a changing world with intellect, humility, and previously unheard-of accuracy.

The groundwork has been established in this chapter for comprehending how AI can transform climate modelling from a static analytical tool into a dynamic, learning system. Building a multifaceted picture of how AI is turning into a catalyst for climate resilience across many spheres of society, we will examine in the upcoming chapters how these insights translate into disaster response, energy transitions, urban infrastructure, and beyond.

References

Allam, Z., & Cheshmehzangi, A. (2024). The green new deal as a pathway for climate and social justice. In *Sustainable Futures and Green New Deals* (pp. 105–133). Springer Nature Switzerland.

Alshahrani, A. (2025). Bridging Cities and Citizens with Generative AI: Public Readiness and Trust in Urban Planning. *Buildings, 15*(14), 2494.

Cash, B. A., Schneider, E. K., & Bengtsson, L. (2005). Origin of regional climate differences: Role of boundary conditions and model formulation in two GCMs. *Climate Dynamics, 25*(7), 709–723.

Cheshmehzangi, A. (2026). *Generative AI-Powered Urban Digital Twins: Pioneering Environmental Solutions for Sustainable Intelligent Cities*. Springer.

Chokkavarapu, N., & Mandla, V. R. (2019). Comparative study of GCMs, RCMs, downscaling and hydrological models: A review toward future climate change impact estimation. *SN Applied Sciences, 1*(12), 1698.

Djouama, I., Kadache, N., & Seghir, R. (2024, September). ML-based weather forecasting models: A comparative study. In *Novel & Intelligent Digital Systems Conferences* (pp. 393–404). Springer Nature Switzerland.

Gopalakrishnan, T., Hasan, M. K., Haque, A. S., Jayasinghe, S. L., & Kumar, L. (2019). Sustainability of coastal agriculture under climate change. *Sustainability, 11*(24), 7200.

Kong, D., Cheshmehzangi, A., Zhang, Z., Ardakani, S. P., & Gu, T. (2023). Urban building energy modeling (UBEM): A systematic review of challenges and opportunities. *Energy Efficiency, 16*(6), 69.

Mizielinski, M. S., Roberts, M. J., Vidale, P. L., Schiemann, R., Demory, M. E., Strachan, J., Edwards, T., Stephens, A., Lawrence, B. N., Pritchard, M., & Chiu, P. (2014). High-resolution global climate modelling: The UPSCALE project, a large-simulation campaign. *Geoscientific Model Development, 7*(4), 1629–1640.

Murukesh, M., Golla, S., & Kumar, P. (2024). Downscaling and reconstruction of high-resolution gridded rainfall data over India using deep learning-based generative adversarial network. *Modeling Earth Systems and Environment, 10*(2), 2221–2237.

Ojanen, A. (2025). Comparative analysis of long-term governance problems: Risks of climate change and artificial intelligence. *Futures & Foresight Science, 7*(1), Article e203.

Rehan, H. (2022). Enhancing Disaster Response Systems: Predicting and Mitigating the Impact of Natural Disasters Using AI. *Journal of Artificial Intelligence Research, 2*(1), 501.

Sathvara, P. K. B., Anuradha, J., Tripathi, S., & Sanjeevi, R. (2024). Prediction Models for Climate Change Using Artificial Intelligence. In *Artificial Intelligence Techniques for Sustainable Development* (pp. 359–379). CRC Press.

Thekdi, S., Tatar, U., Santos, J., & Chatterjee, S. (2023). Disaster risk and artificial intelligence: A framework to characterize conceptual synergies and future opportunities. *Risk Analysis, 43*(8), 1641–1656.

Vahedi, S., Zhao, J., Pierre, B., Lei, F., Anagnostou, E., He, K., Jones, C., & Wang, B. (2025). Wildfire and power grid nexus in a changing climate. *Nature Reviews Electrical Engineering*, 1–19.

Chapter 3
From Signal to Survival: AI-Driven Early Warning Systems for a Climate-Risk Era

3.1 The Evolution of Early Warning Systems: From Reactive to Predictive

For generations, Early warning systems (EWS) have been the first line of defence for society against natural dangers (Quansah et al., 2010). These systems have attempted to foresee environmental risks and initiate prompt actions, from the employment of meteorological radar in modern times to the river gauge readings made by ancient civilizations. But despite their significance, traditional EWS have always had serious drawbacks, chief among them being short lead periods, data silos, and poor accuracy in rapidly changing climate scenarios.

Early warning systems have traditionally depended on deterministic thresholds, such as drought declarations based on months of rainfall shortfall, cyclone alerts sent whenever a storm achieves a specific wind speed, or flood warnings based on river height surpassing a crucial level. As a result, they have turned to threshold-based, deterministic methods that are based on empirical data. To identify when an event crosses a predetermined danger point, meteorological authorities track factors including river heights, wind speeds, and rainfall intensity (Wang et al., 2017). Because they are based on observed occurrences that are currently happening, these signals are frequently reactive. Because of this, warnings usually occur late in the hazard lifetime, providing little time for evacuation or preparation. The problem is exacerbated for slow-onset calamities, such as droughts, by trailing indicators that fail to detect early changes in soil moisture or weather patterns. The difference between containment and disaster for sudden-onset disasters, such as cyclones, flash floods, or wildfires, can be determined by the timing of warnings.

Furthermore, a lot of conventional EWS operate in geographical and sectoral silos. Hydrological, ecological, and social datasets may be handled independently from meteorological data (Twigg, 2021). Systems used to notify coastal metropolitan dwellers may differ greatly from those used to notify farmers in rural areas. In addition to impeding a coordinated response, this fragmentation restricts the warnings'

© The Author(s), under exclusive license to Springer Nature Singapore Pte Ltd. 2025

A. Cheshmehzangi, *AI Agency in Climate Emergency*, Urban Sustainability,
https://doi.org/10.1007/978-981-95-4192-8_3

capacity to react to complex, compound hazards, such as a drought that increases the risk of wildfires or a cyclone that is followed by flash flooding.

Such restrictions are unacceptable in the age of climate instability. The unpredictability, intensity, and frequency of disasters are increasing. The courses of storms are changing in real time. Fires spread quickly across arid terrain. With static models, precipitation extremes are becoming increasingly hyper-localized and challenging to predict. Gaps in coverage and communication infrastructure put vulnerable communities, who frequently reside in marginal lands, informal settlements, or areas with limited access to the internet, at disproportionate danger.

A paradigm change from reactive warnings to predictive, adaptive, and integrated early warning systems is required due to the increasing urgency. AI offers a significant advancement in this area. AI is capable of processing massive amounts of data in real time from sensors, satellites, weather models, and even crowdsourced inputs. It can quickly and precisely replicate changing risks, identify subtle patterns, and spot anomalies. More significantly, AI-driven systems are dynamic; unlike traditional systems, they can localize predictions, adapt dynamically to changing inputs, and learn from past events.

A new type of warning system is created, one that is context-sensitive rather than generic, transdisciplinary rather than compartmentalized, and anticipatory rather than reactive. These AI-powered early warning systems provide more than just a warning that a threat is approaching; they also provide personalized warnings, actionable insights, and nuanced foresight that may be implemented well in advance of a tragedy. The next part examines the technological and practical ways in which AI accomplishes this, serving as the central nervous system that integrates intricate data, recognizes threats in several fields, and converts signals into intelligence that can save lives.

3.2 AI as the Nerve Center: Data Fusion and Pattern Recognition

The ability to detect, decipher, and predict complex environmental signals in real time is at the core of next-generation early warning systems. AI-powered early warning infrastructures serve as the planet's nervous system, receiving constant streams of environmental data, identifying emerging threats through pattern recognition, and quickly producing actionable forecasts. This is in contrast to traditional systems that rely on fixed thresholds and discrete data inputs. This change is revolutionary rather than gradual. AI has emerged as the engine that combines disparate data and transforms it into high-precision warnings with previously unheard-of speed and accuracy, especially through machine learning and deep learning.

The ability of AI to combine diverse data sources, such as satellite imagery, radar systems, IoT-enabled sensors, weather station feeds, terrain elevation maps, and even unconventional inputs like social media posts, emergency call logs, and drone footage, is one of its most alluring features (Cheshmehzangi, 2025). Because of scale, format, or latency incompatibilities, traditional models frequently isolate these streams. AI, on the other hand, benefits greatly from high-dimensional data. AI systems may combine structured and unstructured inputs into cohesive predictive models by using methods like data fusion, neural embedding, and feature extraction. This makes it possible to find hidden connections and weak signals that would otherwise be undetectable with traditional techniques.

For instance, computer vision and deep learning have become effective methods for detecting early fire ignition and patterns of fire spread in wildfire detection. In order to identify anomalies like smoke plumes, hot patches, and changes in the greenness of the vegetation, convolutional neural networks (CNNs) are trained to examine satellite thermal imaging, multispectral inputs, and near-infrared data. In addition to determining whether a fire has begun, AI can determine its rate of spread, direction, and topography. By providing predictive insights, this technology enables responders to take action before the situation gets out of control. AI models frequently perform hours or even days better than human detection, greatly enhancing containment results.

Similarly, the application of spatiotemporal models like Long Short-Term Memory (LSTM) networks and convolutional LSTMs has led to significant advancements in AI-enhanced flood forecasting (Liu et al., 2024). Both temporal dynamics, such as rainfall progression and river flow velocity, and spatial interactions, such as land use and terrain slope, are captured by these models. AI systems, in contrast to static hydrological models, are able to adapt dynamically to changing catchment saturation levels, incoming precipitation data, and infrastructural limitations such as dam operations or drainage capacity. This makes it possible to anticipate urban floods in real time, down to the block or street level, which is essential for highly populated areas that are vulnerable to flash flooding. Mobility data is also used by some systems to optimize evacuation routes and emergency traffic flow.

The ability of AI to monitor small ecosystem changes over time is especially beneficial for drought monitoring, a more gradual and spatially dispersed threat. To find early indicators of agricultural stress and ecological deterioration, anomaly detection algorithms examine remote sensing inputs, including NDVI (Normalized Difference Vegetation Index), surface soil moisture levels, evapotranspiration rates, and temperature anomalies (Joiner et al., 2018). Autoencoders and clustering algorithms are two examples of unsupervised machine learning approaches that aid in identifying environmental pattern deviations that occur prior to drought conditions. These models provide more fair access to predictive information, making them especially useful for areas with limited or delayed traditional meteorological coverage.

AI-powered ensemble prediction systems are improving the accuracy and speed of storm tracking for hurricanes and cyclones. AI systems can improve trajectory forecasts and more accurately estimate impact zones by combining outputs from several numerical weather models with satellite and radar data (Nardi et al., 2025). To provide more localized route and intensity estimates, recurrent neural networks and transformer-based designs are skilled at learning from historical cyclone behaviour.

They do this by considering factors such as wind shear, ocean heat content, atmospheric pressure patterns, and coastline terrain. This enhances the lead time for storm surge warnings and aids in directing prompt evacuation orders, particularly in coastal areas with high population density.

The three main performance improvements that AI-enhanced early warning systems provide across these threat classes are as follows (Fig. 3.1):

Faster detection—Machine learning models operate at speeds orders of magnitude faster than human analysts or physics-based simulations, enabling near real-time risk identification.

More accurate risk mapping—AI systems pinpoint high-risk zones with far greater spatial precision, taking into account localized variables and interdependencies.

Extended lead times—By predicting events before threshold breaches occur, AI systems buy valuable time for preventive measures and emergency coordination.

This predictive advantage is not just technical; it can save lives. AI enables organizations and communities to react proactively rather than reactively, whether that

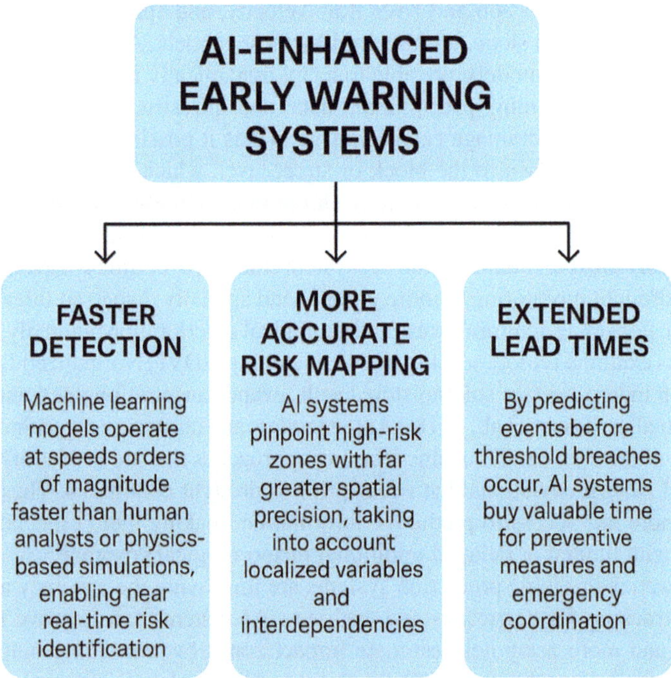

Fig. 3.1 Three main performance improvements for AI-enhanced early warning systems

means spotting slow-building drought indicators weeks before agricultural failure, anticipating urban inundation hours in advance, or identifying a wildfire in its ignition phase. In sum, AI is the brains behind a new type of early warning system that can see across several domains, reason in real time, and provide humans with previously unheard-of clarity when making decisions. However, in order to achieve this potential, we also need to think about how these signals are conveyed, understood, and responded to—a topic covered in the following section.

3.3 Alerts that Matter: Localized, Interpretable, and Actionable

Any early warning system's ability to alert and safeguard individuals in danger ultimately determines its efficacy, not its level of sophistication. No matter how technically correct, an alert is only as good as its target audience if it is timely, understandable, and actionable. The speed and granularity of threat identification are increased by AI-powered systems, but there is a concomitant difficulty of making sure that alarms are sent to the appropriate individuals at the appropriate time. This calls for a shift away from crude predictions and toward impact-based, user-centered communication frameworks, an area in which AI is becoming increasingly important.

The transition from event-based forecasting, such as "heavy rainfall expected," to impact-based forecasting, such as "expect flooding of low-lying streets" or "power outages likely in eastern districts," is one of the most important changes made possible by AI. For instance, AI systems can model not only environmental change but also its ripple effects on services, infrastructure, human health, and mobility. Through the integration of exposure and vulnerability models with hazard data, AI also facilitates multi-layered evaluations of possible outcomes. With this method, warnings are converted from general meteorological data into specific, regional instructions for the people who need to react (Fig. 3.2).

In this situation, tailoring is necessary—not optional. Farmers must be informed in advance of the likelihood of a pest outbreak or drought, preferably by SMS notifications in their native tongues. Real-time maps that display at-risk populations, road accessibility, and fire perimeters are essential for first responders. Mobile push notifications that provide information on flood routes or heat-related safety measures may be helpful to urban dwellers. Policymakers need scenario briefings backed by dynamic models that include estimates of the socioeconomic impact. These distinct interfaces are made possible by AI, which modifies alert content and delivery based on audience demands, digital access, and cognitive load.

Here, Natural Language Generation (NLG) models are crucial. AI increases the accessibility of early warning systems across languages, cultures, and literacy levels by converting numerical outputs or geospatial data into understandable, context-sensitive, multilingual messages (Chakrabarty et al., 2023). Chatbots and virtual assistants, educated on local vernaculars and linked to hazard models, can

Fig. 3.2 The transition from event-based forecasting with AI use and integration

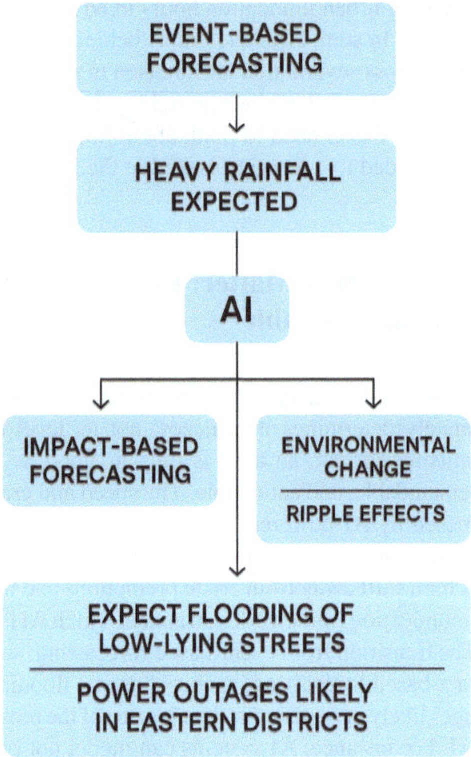

provide two-way communication platforms in linguistically diverse places, enabling users to report circumstances in real time, ask questions about danger levels, and seek safety recommendations. Pilot programs for these technologies are currently underway in disaster-prone areas with high mobile penetration and little government infrastructure.

Visualization also matters. Both experts and non-experts can engage with real-time data in straightforward ways with the help of geospatial dashboards and digital twin platforms supplemented with AI overlays (Abdeen, 2025; Pourroostaei Ardakani & Cheshmehzangi, 2024). Users can view real-time floodplains, heat zones, or cyclone routes along with suggested courses of action. By filtering data based on user profiles, locations, and past risk encounters, AI personalizes these interfaces. Predictive maps with asset and demographic data superimposed provide quick triage of areas that require the most action for institutional customers.

However, as our reliance on AI increases, explainability becomes even more crucial. Models must not only generate accurate predictions but also offer clear reasoning in order to increase public officials' and emergency agencies' confidence in

AI-driven notifications. For what reason was a particular neighbourhood designated as high-risk? Which factors influenced the model's decision the most? By using methods like SHAP values, attention mechanisms, or rule-based hybrid systems, interpretable AI makes sure that alerts are not "black boxes," but rather reliable instruments that can be examined, verified, and enhanced over time.

Establishing this trust is essential. Without it, important stakeholders could be reluctant to respond to warnings from AI, which could cause delays or poor management. Furthermore, excessively frequent, ambiguous, or false signals run the risk of desensitizing the public, a condition known as warning fatigue. In order to stay accurate and relevant over time, AI systems must be built with social reliability in addition to technical performance, incorporating human input and local expertise.

In sum, alerts that empower are the ones that count. AI turns abstract data into intelligence that can save lives by localizing warnings, converting them into actionable formats, and establishing their foundation in local reality. However, in order to realize this promise, early warning systems need to give equal weight to communication, context, and clarity as they do to computation—a fine balance that can only be reached by fully human-centered AI. In order to make AI-powered early warning systems not only intelligent but also just, the last section examines how these technologies can be expanded fairly across populations and geographical areas.

3.4 Equity, Ethics, and Global Reach: Making AI Early Warnings Inclusive

A crucial point that needs to be addressed as AI-powered early warning systems advance in capability and sophistication is who gains and who loses out. AI's disruptive potential cannot be completely realized until it takes into account social justice, geographic inclusivity, ethical accountability, and technical efficiency. Due to data scarcity, a lack of digital infrastructure, and brittle communication systems, the basic underpinnings of AI deployment remain precarious in many parts of the world, especially in low-income, rural, or conflict-affected areas.

Even the most accurate models are useless in underserved areas if populations do not have access to smartphones, the internet, or reliable institutions to act on warnings. If AI is to make a significant contribution to global resilience, it is imperative that underlying structural disparities—economic, educational, and linguistic—be directly addressed, as reflected in this digital divide. AI models run the danger of functioning in blind spots, which could reinforce exclusionary behaviours or completely misrepresent threats in areas with limited satellite data, no local weather stations, or poor sensor networks.

However, there are some exciting new developments. For instance, millions of people in low-lying areas of Bangladesh are now receiving SMS-based alerts about impending floods because to AI-enhanced flood forecasting models that were created through partnerships between Google, the government, and non-governmental organizations. AI and remote sensing are being utilized in sub-Saharan Africa to pinpoint areas that are vulnerable to wildfires and notify rural residents via offline apps and community radio. These initiatives show that AI may be modified to function under limitations rather than only in data-rich settings with the correct collaborations.

However, inclusion involves more than just access; it also involves accountability and agency. Systems powered by AI must be built to prevent bias in the prioritization of certain populations and the assessment of dangers. False negatives, or missed threats, can be disastrous, while false positives, or alerts that never come to pass, can damage trust. Relying too much on opaque algorithms could undermine frontline responders or circumvent local expertise. A burgeoning movement is advocating for open-source platforms, civic alliances, and locally co-designed technologies that enable communities to engage in the creation, calibration, and interpretation of models in order to reduce these risks.

According to this paradigm, AI enhances human judgment rather than replaces it. Instead of acting as a top-down oracle, it turns into a partner in readiness. For early warning systems to benefit the public rather than just maximize performance metrics, ethical AI frameworks—such as transparency norms and feedback mechanisms—are crucial. In conclusion, AI has much more potential for early warning systems than just producing more accurate maps or quicker forecasts. It resides in its capacity to democratize foresight, equip marginalized populations with the information they need to take action before a tragedy occurs, and create a future in which being prepared is a shared right rather than a privilege. AI will not only make warnings more intelligent, but also more just if we incorporate ethics and equality into the core of these systems.

References

Abdeen, N. (2025). *Call for Citizen-Centric Digital Twins for Transport Infrastructure* (Doctoral dissertation, UNSW Sydney).

Chakrabarty, T., Padmakumar, V., He, H., & Peng, N. (2023, December). Creative natural language generation. In *Proceedings of the 2023 Conference on Empirical Methods in Natural Language Processing: Tutorial Abstracts* (pp. 34–40).

Cheshmehzangi, A. (2025). *Urban Innovation and Sustainability: ICT-based Interventions in Hainan, China.* Springer Nature.

Joiner, J., Yoshida, Y., Anderson, M., Holmes, T., Hain, C., Reichle, R., Koster, R., Middleton, E., & Zeng, F. W. (2018). Global relationships among traditional reflectance vegetation indices (NDVI and NDII), evapotranspiration (ET), and soil moisture variability on weekly timescales. *Remote Sensing of Environment, 219*, 339–352.

Liu, Z., Coleman, N., Patrascu, F. I., Yin, K., Li, X., & Mostafavi, A. (2024). Artificial intelligence for flood risk management: A comprehensive state-of-the-art review and future directions. *International Journal of Disaster Risk Reduction*, 105110.

Nardi, E., Cohen, O., Pinhasi, Y., Haridim, M., & Gavan, J. (2025). Weather Forecasting Satellites—Past, Present, & Future. *Information, 16*(8), 677.

Pourroostaei Ardakani, S., & Cheshmehzangi, A. (Eds.). (2024). *Digital twin computing for urban Intelligence.* Springer.

Quansah, J. E., Engel, B., & Rochon, G. L. (2010). Early warning systems: A review. *Journal of Terrestrial Observation, 2*(2), 5.
Twigg, J. (2021). *Sustainability of early warning systems.* Discussion Paper. Building Resilience and Adapting to Climate Change (BRACC). Discussion Paper, February 2021, 29 pages.
Wang, X., Wang, S., Ren, L., & Zeng, Z. (2017). Spatial distribution of rainstorm hazard risk based on EW-AHP in mountainous scenic area of China. *Human and Ecological Risk Assessment: An International Journal, 23*(4), 925–943.

Chapter 4
Seconds Ahead: AI for Real-Time Disaster Response and Life-Saving Logistics

4.1 The High-Stakes Window: Challenges of Disaster Response in a Climate Era

The difference between life and death can be made in the crucial hours after a calamity. Often referred to as the "golden hours" in disaster response, this short and unstable period is when coordinated action is most important, i.e. when it is necessary to rescue injured individuals, distribute critical resources, and stop cascading failures. However, this is also the time when there is the greatest amount of uncertainty: communications may be cut off, infrastructure may be unavailable, and ground intelligence may be scant or out-of-date. The effectiveness of reaction systems is put to the test in these high-stakes situations where informational chaos and time sensitivity collide.

First responders always encounter operational difficulties that prevent prompt and efficient action all over the world. Fire departments, ambulance services, crisis management organizations, and non-governmental organizations frequently communicate in fragmented ways using incompatible systems or unofficial routes. Limited data, sensor blackouts, or physical inaccessibility hinder real-time visibility into disaster areas (Daud et al., 2025). Incomplete or delayed information is often used to make decisions on whether to order evacuations, where to deploy relief, and how to prioritize areas, running the danger of both overreach and important omissions. This is made much more difficult by the logistical challenges of rerouting emergency vehicles, negotiating damaged infrastructure, and assigning limited resources under extreme strain.

These problems are getting worse due to climate change. There is no longer any isolation, seasonality, or predictability in disasters. The frequency and complexity of compound occurrences are increasing, such as tropical cyclones causing landslides and urban flooding or heatwaves starting wildfires during drought. These incidents frequently have a cascading effect on systems; for example, a flood can paralyze transportation, disrupt telecommunications, and shut down electrical infrastructures

© The Author(s), under exclusive license to Springer Nature Singapore Pte Ltd. 2025
A. Cheshmehzangi, *AI Agency in Climate Emergency*, Urban Sustainability,
https://doi.org/10.1007/978-981-95-4192-8_4

in a matter of hours. As a result, disaster response now functions in a more complex environment where events are becoming less linear, faster-paced, and more spatially and temporally overlapping.

Urbanization increases stress levels. Rapid, fair catastrophe response is made more difficult in megacities by the sheer density and diversity of the population, infrastructure risks, and socioeconomic inequities. Problems are frequently the opposite in rural or remote areas, where inadequate infrastructure, bad connectivity, and a lack of institutional presence put communities at risk of being dangerously isolated during emergencies.

Predictive, adaptive, and real-time solutions that can help emergency services navigate uncertainty are desperately needed to fulfill these growing demands. Delays in reporting and static procedures are no longer adequate. The transition from reactive to anticipatory coordination is necessary so that information may be collected, analyzed, and taken action on as a disaster occurs, rather than hours or days later.

Here is where the transformational power of AI is beginning to emerge. AI systems can produce dynamic maps of events as they happen by consuming enormous volumes of organized and unstructured data, such as satellite feeds and crowdsourced alerts (Alamri, 2024). They are able to identify the areas with the highest requirements, the routes that are still feasible, and the minute-by-minute changes in dangers. In addition to helping rescuers comprehend what is happening, AI also helps them predict what will happen next, allowing for quicker, more focused, and more flexible actions.

By mapping at-risk groups, evaluating ground-level data in real-time, and giving rescuers the intelligence they need to make quick decisions under duress, the next section examines how AI improves situational awareness.

4.2 Eyes on the Ground: AI for Risk Detection, Population Mapping, and Vulnerability Assessment

Knowing what is happening and to whom is the most precious money in the pandemonium of a disaster. However, this information is frequently buried beneath damaged infrastructure, communication connections are jammed, or it is developing too quickly for human operators to decipher. This vacuum is being filled by AI, which gives rescuers what they most need: eyes on the ground, everywhere, at once.

An array of high-frequency data streams can be quickly processed and synthesized by AI-powered systems to produce a current and cohesive image of crisis situations (Cheshmehzangi, 2026; Kumar et al., 2024). Computer vision algorithms use satellite data, drone footage, and radar scans to detect landslide scars, flooded communities, collapsed structures, and wildfire perimeters, often within minutes of the image being taken. As the scenario changes, these systems continuously update threat borders, highlight damage areas, and detect changes versus baseline imagery. These tools are essential for safe travel and response priority in areas of conflict or earthquake aftermath when ground access is hazardous or unavailable.

Simultaneously, social media analysis has emerged as a surprising yet effective catastrophe intelligence tool. Natural language processing-trained AI models can keep an eye out for distress messages, image uploads, or keywords associated with certain areas on sites like Facebook, X (previously Twitter), and neighbourhood apps (Kumar et al., 2023). After being geotagged and filtered, these signals help create real-time "crisis heat maps" that show confirmed incident reports, public opinion, and emergency needs. Crucially, AI can discriminate between legitimate alarms and noise, allowing authorities to prioritize locations according to accessibility, severity, and immediacy (See Fig. 4.1).

Population vulnerability assessment is a crucial component of situational awareness afforded by AI. Even before a crisis occurs, AI models can identify high-risk groups, such as the elderly, disabled, children, the chronically ill, and marginalized communities, by combining census data, geographical records, urban mobility data, and demographic profiles. These models take into account both dynamic exposure

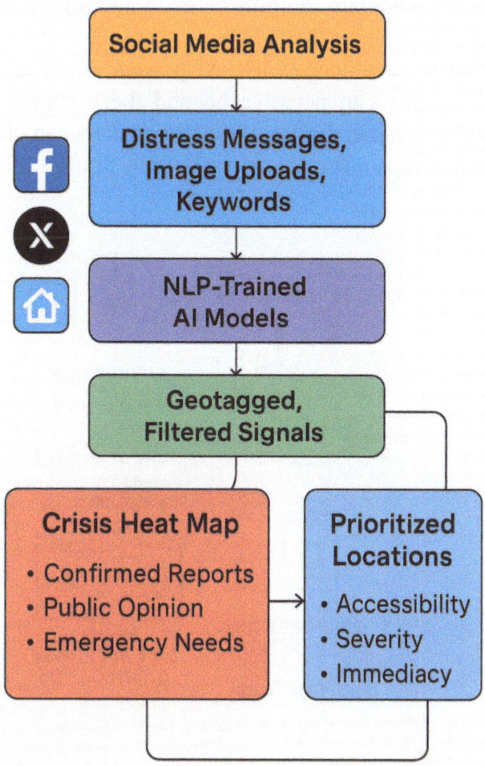

Fig. 4.1 AI-powered social media analysis transforms raw distress signals into actionable crisis intelligence. Posts, images, and keywords from platforms like Facebook, X, and neighbourhood apps are processed by NLP models, geotagged, and filtered to create real-time crisis heat maps. Authorities can then prioritize locations by accessibility, severity, and urgency for targeted emergency response

(such mobility restrictions or actual proximity to hazard zones) and static vulnerability (like income level or housing quality). AI systems can identify not only where individuals are but also who needs help the most urgently by comparing these layers.

AI-assisted satellite mapping technologies, for example, were used in a matter of hours during the 2023 earthquakes in Turkey to determine collapsed areas, calculate structural damage, and establish a relationship between building density and population exposure. These revelations helped rescue crews identify the most critical search and recovery locations and guided international aid flows to the most affected towns. Similar to this, in California, evacuation assistance for elderly and medically vulnerable citizens during wildfire situations has been prioritized using AI-driven vulnerability scoring models that combine fire weather forecasts with information on age, health, and mobility.

Another new area is tracking crowd density in real time. AI systems can identify unusual aggregation patterns, halted evacuations, or roads that are getting crowded or inaccessible by evaluating aggregated and anonymized mobile phone geolocation data (Abraham et al., 2024). Live response dashboards can include this data to reroute traffic, modify resource allocation, or send out stranding notifications. When there are limited or transient escape routes due to floods, heat waves, or chemical releases, this is particularly helpful in urban environments for controlling mass movements (See Fig. 4.2).

When combined, these AI powers create a multi-layered, adaptive awareness system that can perceive beyond the obvious, deduce patterns of urgency, and highlight vulnerabilities. Decision-makers can obtain dynamic, continually updated insights that inform decisions hour by hour as crises develop, rather than depending

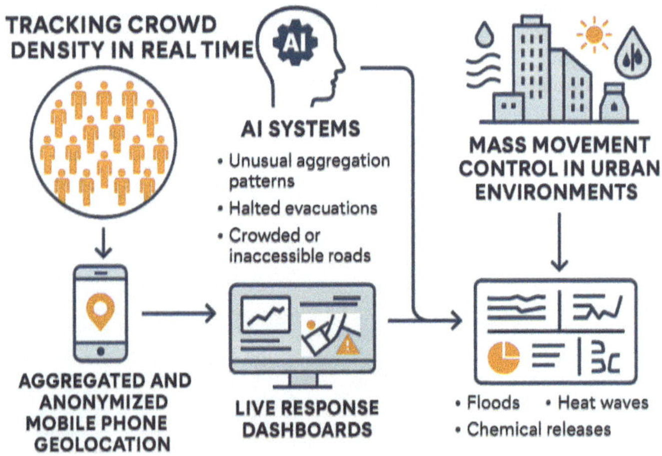

Fig. 4.2 Summary of tracking crowd density in real time using AI systems for optimisation and mass movement control in urban environments

only on static risk maps or post-event assessments (Wright, 2023). However, intelligence is only as good as the action it makes possible. The following part examines how AI transforms detection into a life-saving intervention by assisting in the timely and efficient movement of medical, logistical, and human resources.

4.3 Machines that Move with Purpose: AI for Logistics, Routing, and Resource Optimization

Not only must communication flow quickly during a disaster, but resources also need to move quickly. Despite damaged roads, interrupted supply chains, or unpredictable weather, food, fuel, water, medication, generators, and rescue personnel must quickly reach affected communities. Under these circumstances, the margin of error drastically reduces. Delivery delays may result in fatalities. Misallocation of resources can make suffering worse. Here is where AI truly shines—not only in identifying demands, but in coordinating the movement of all the important things in real time.

Supply chain optimization driven by AI is at the core of this capability. AI systems can optimize the scheduling and routing of emergency supplies by using real-time stocks, road conditions, hazard forecasts, and past disaster data (Abaku et al., 2024; Grover, 2025). Machine learning techniques predict the areas with the most demand and recommend the best distribution plans. AI can, for instance, dynamically prioritize delivery from neighbouring depots in high-density shelter zones with severe clean water shortages in post-earthquake scenarios. It can reroute oxygen tanks or medical supplies from dangerous paths to safer corridors during quickly changing wildfire outbreaks (See Fig. 4.3).

These systems depend on ongoing recalibration and the deep integration of geospatial, logistical, and environmental data sources. AI instantly modifies the distribution strategy in the event of a bridge collapse, a supply vehicle delay, or the emergence of a new danger zone (Tang & Huang, 2024). The ability to react and adjust to interruption rather than only predict it is what sets AI-enabled logistics apart from static contingency models.

Dynamic routing for emergency response vehicles is equally important. Road accessibility, traffic, or weather can seriously restrict progress for rescue teams, ambulances, and fire vehicles, who operate under tremendous time constraint. To determine the quickest and safest routes across chaotic surroundings, AI systems use real-time traffic data, GPS feeds, blockage reports, and even crowdsourced updates. Such routing intelligence can be the difference between survival and death when every minute matters.

More sophisticated implementations make use of multi-agent systems and reinforcement learning, in which AI agents simulate innumerable encounters in crisis scenarios to determine the best course of action. For instance, one agent could stand in for a drone network that relays information, another for a convoy of supply trucks, and a third for a rescue aircraft. These systems test scenarios, model coordination

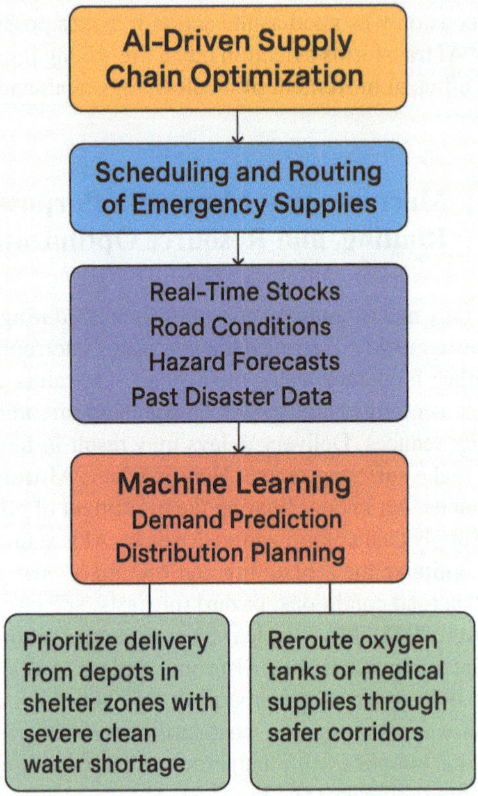

Fig. 4.3 AI-driven supply chain optimization enhances emergency response. Real-time data on stocks, road conditions, hazard forecasts, and historical disasters feed into machine learning models for demand prediction and distribution planning. The system dynamically prioritizes deliveries, such as sending water to shelter zones or rerouting medical supplies through safer paths during evolving crises

patterns, and improve teamwork tactics according to predetermined objectives (e.g., avoiding bottlenecks, maximizing coverage, or lowering delivery time). This kind of learning is perfect for high-complexity, multi-scale activities since it is dynamic and not pre-scripted.

Proactive scenario simulations and evacuation modelling are two of the most promising applications of AI-driven logistics. By detecting possible chokepoints, estimating clearance durations, and recommending staged evacuations depending on population density and transportation availability, AI can model how various hazard scenarios may affect mobility throughout a city or area. Such modelling assists proactive decisions on who should evacuate first, where to send them, and how to communicate most effectively in areas with limited infrastructure or time-sensitive dangers (such as flash floods or tsunami zones).

It goes a step further with predictive resource deployment. AI systems use demographic trends, seasonal risk models, and past disaster patterns to suggest pre-positioning supplies and equipment well in advance of a crisis. For example, AI models may recommend that coastal communities in Southeast Asia's cyclone-prone areas stock up on food, fuel, and satellite phones prior to the official alert period. This proactive measure not only speeds up reaction times but also strengthens the system's resilience, enabling relief efforts to start even when communications are partially interrupted.

Applications in the real world validate this potential. AI-enabled cyclone evacuation systems, created in partnership with state and federal agencies, have significantly decreased the number of cyclone-related fatalities in Odisha, India, in recent years. To more effectively direct mass evacuations, these models integrate real-time vulnerability evaluations, movement tracking, and weather forecasts. AI-assisted logistics systems were used after Hurricane Maria in Puerto Rico to map remote communities, forecast areas where relief was most needed, and direct trucks and helicopters across difficult terrain. In both situations, AI allowed responders to outpace chaos with strategic, flexible response in addition to streamlining logistics.

Finally, disaster response logistics are an exercise in coordination under duress. Decisions must be taken fast, resources must be allocated wisely, and responders must act with little visibility during a crisis. By coordinating the moving components of an emergency response system with vision, adaptability, and focus, AI gives coherence to this complexity. As AI gains operational power, the last section considers how it must continue to be inclusive and accountable, focusing on human-AI cooperation, moral protections, and fair results.

4.4 Coordinated Intelligence: Human-AI Collaboration and Ethical Considerations in Disaster Response

In the urgency of disaster response, the pressure to act swiftly can often blur the boundaries between human judgment and machine-generated recommendations (He & Fang, 2024). However, AI must continue to be a collaborator, not a replacement, as it becomes more integrated into emergency operations, from locating hotspots to allocating resources. Coordinated intelligence, where AI augments human responders' skills without superseding their expertise, instinct, or responsibility, is the way of the future for efficient disaster response, not automated systems operating in silos.

The idea of human-in-the-loop systems is at the heart of this synergy. AI can identify patterns, filter out noise, and recommend the best course of action, but trained people should always make the final decisions, especially when it comes to human life. Responders must have faith in an AI's accuracy, transparency, and interpretability in high-stakes situations where seconds matter. A system must be able to provide an explanation for its recommendations, such as which areas should be evacuated first or which supply routes should be prioritized. Models that lack logic or are opaque run the risk of being disregarded, or worse, mindlessly adopted without question. Therefore, explainability is a moral and practical requirement rather than a technical luxury. It makes it possible for government representatives, emergency coordinators, and medical professionals to comprehend the reasoning behind AI insights, evaluate dangers suitably, and incorporate those insights into more comprehensive reaction plans. Under pressure, when speed is just as important as clarity, this becomes much more crucial.

At the same time, there are serious ethical concerns with the expanding use of AI in emergency situations, especially with regard to data use. Sensitive data is necessary for real-time population mapping, drone or smartphone monitoring, and predictive profiling of vulnerable populations. Such tools run the danger of infringing civil rights, escalating social biases, or unduly monitoring vulnerable communities in the absence of strong data governance, privacy protections, and informed consent procedures. People's liberty and dignity must be upheld, even during emergencies.

AI-enabled disaster response must be inclusive by design in order to live up to its potential (Cheshmehzangi, 2025). This entails making certain that those who are frequently overlooked—undocumented people, those with disabilities, non-digital users, the homeless, or those living in informal settlements—are reached by notifications, resources, and evacuation assistance. To ensure that no one is left out of the safety net, delivery methods must be adapted to numerous communication modalities, including radio, SMS, multilingual signage, and local intermediaries. AI models must also be trained on a variety of datasets that represent inequalities in the real world.

The ultimate goal of AI in disaster response is to increase responders' efficacy, reach, and foresight when every second counts, not to replace them (Cheshmehzangi & Pourroostaei Ardakani, 2024). AI becomes more than just a tool when it is developed morally, used sensibly, and cooperatively integrated. It becomes an ally in the struggle to save lives in the most trying circumstances. This chapter has demonstrated that AI's ability to make complexity actionable is just as powerful in disaster response as its computational prowess. AI makes it possible for more quick, accurate, and scalable reactions in a variety of areas, including logistics, risk mapping, evacuation planning, and detection. However, in order to fulfill its promise, it must continue to be based on human values, governed by moral principles, and conceived with inclusivity at its center. Artificial and human intelligence must cooperate in times of crisis.

References

Abaku, E. A., Edunjobi, T. E., & Odimarha, A. C. (2024). Theoretical approaches to AI in supply chain optimization: Pathways to efficiency and resilience. *International Journal of Science and Technology Research Archive, 6*(1), 092–107.

Abraham, A., Zhang, Y., & Prasad, S. (2024). Evacuation management framework towards smart city-wide intelligent emergency interactive response system. *arXiv preprint* arXiv:2403.07003

Alamri, S. (2024). The geospatial crowd: Emerging trends and challenges in crowdsourced spatial analytics. *ISPRS International Journal of Geo-Information, 13*(6), 168.

Cheshmehzangi, A. (2025). Bridging Innovation and Sustainability: Global Insights on ICT Applications. In *Urban Innovation and Sustainability: ICT-based Interventions in Hainan, China* (pp. 23–43). Springer Nature Singapore.

Cheshmehzangi, A. (2026). *Generative AI-Powered Urban Digital Twins: Pioneering Environmental Solutions for Sustainable Intelligent Cities*. Springer.

Cheshmehzangi, A., & Pourroostaei Ardakani, S. (2024). Embracing the digital twin paradigm for urban sustainability. In *Digital Twin Computing for Urban Intelligence* (pp. 1–11). Springer Nature Singapore.

Daud, A., Al Abdouli, K. M., & Badshah, A. (2025). Emerging Computing Tools for Emergency Management: Applications, Limitations and Future Prospects. *IEEE Open Journal of the Computer Society.*

Grover, N. (2025). AI-Enabled Supply Chain Optimization. *International Journal of Advanced Research in Science, Communication and Technology*, 28–44.

He, X., & Fang, L. (2024). Regulatory challenges in synthetic media governance: Policy frameworks for AI-generated content across image, video, and social platforms. *Journal of Robotic Process Automation, AI Integration, and Workflow Optimization, 9*(12), 36–54.

Kumar, R. P., Muthukumaran, N., Hanisha, C., Rithvik, P., & Goud, M. S. (2023, October). Examination system automation using natural language processing. In *2023 International Conference on Self Sustainable Artificial Intelligence Systems (ICSSAS)* (pp. 1002–1008). IEEE.

Kumar, Y., Marchena, J., Awlla, A. H., Li, J. J., & Abdalla, H. B. (2024). The AI-powered evolution of big data. *Applied Sciences, 14*(22), 10176.

Tang, D., & Huang, M. (2024). The sustainable development of bridges in China: Collapse cause analysis, existing management dilemmas and potential solutions. *Buildings, 14*(2), 419.

Wright, B. (2023). *Resilient decisions: Crisis-event decision making and the influence of post-event scrutiny.* [Doctoral Thesis, Charles Sturt University]. Charles Sturt University.

Part II
AI-Driven Energy System Optimization and Decarbonization

This section focuses on how AI is changing energy systems to improve efficiency and decarbonization. AI-powered smart grids seamlessly integrate renewable energy sources like solar and wind by balancing real-time energy supply and demand. Through the optimization of logistics systems, industrial processes, and building operations, AI also significantly increases energy efficiency. AI facilitates the transportation sector's shift away from fossil fuels by managing networks for electric vehicle charging, easing traffic, and optimizing public transportation. When taken as a whole, these uses establish AI as a vital facilitator of future energy systems that are smarter, cleaner, and more sustainable.

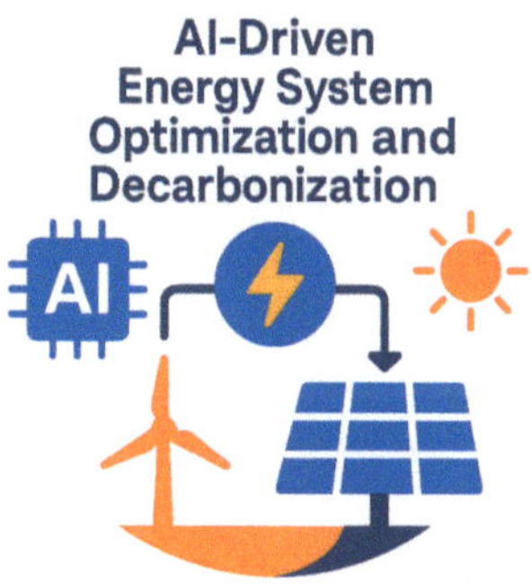

Chapter 5
The Intelligent Grid: AI at the Heart of a Flexible, Renewable Energy Future

5.1 From Rigid to Responsive: The Rise of Smart Grids in the Renewable Era

One of the most significant, although little-known, changes of our time is the development of the electricity grid. Power grids served as the inflexible backbones of industrial society for almost a century. They were built on centralized generation, unidirectional energy flow, and long-range demand forecasting assumptions that rarely bowed but frequently crumbled under pressure. Large power plants, whether coal, gas, hydro, or nuclear, were at the core of these conventional systems. These plants pushed electrons outward through high-voltage transmission lines, which then cascaded down into substations before being distributed to residences, factories, and commercial buildings. This infrastructure, a grid built for predictability in a world assumed to be stable, functioned with a sort of blind regularity. But the conditions that made such a structure possible are no longer the norm (Gilrein et al., 2021). We are now firmly in the age of renewable integration, and the rise of the smart grid marks a turning point from rigidity to responsiveness.

The story of the smart grid is one of conceptual reorientation rather than just technical advancement. The formerly centralized grid started to fragment and diversify as solar panels began to sprout on residential rooftops, wind turbines began to spin across distant hills, and battery systems began to appear in garages and behind utility meters. Power was exchanged, saved, shared, and even forecasted; it was no longer merely created and consumed. In some locations, the energy flow was inverted; in others, it became multidirectional; and in yet others, it changed from static consumption to dynamic interaction. Both digital potential and climatic imperatives played a significant role in this shift. Despite all its environmental benefits, renewable energy brought volatility: wind output varies wildly, solar generation drops off after dusk, and surplus power from prosumers—those who generate and consume—needs new markets and procedures. This unstable, decentralized reality was not suited for the old grid's sluggish decision-making processes and immobile control structures.

© The Author(s), under exclusive license to Springer Nature Singapore Pte Ltd. 2025
A. Cheshmehzangi, *AI Agency in Climate Emergency*, Urban Sustainability,
https://doi.org/10.1007/978-981-95-4192-8_5

More recently, the pressure has increased substantially. Weather anomalies and changing social rhythms produced increasingly sharp and irregular peak demand windows. Mid-twentieth-century infrastructure squeaked under the weight of the twenty-first century. The frequency of outages increased, as did the severe nature of cascade failures. The complexity was further increased by the emergence of smart appliances, microgrids, and electric cars, which produced a flurry of patterns in generation and demand that were impossible to predict linearly. The grid needed to learn to anticipate, adjust, and listen in order to manage this complexity. The intelligent grid era was upon us.

In addition to being improvements to an antiquated system, smart grids were reimaginings of what a grid could be: intelligent, decentralized, interactive, and self-regulating. These grids work more like living systems than mechanical ones because they are outfitted with sensors, edge devices, real-time communication protocols, and sophisticated control systems. They take into account input from rooftop solar arrays, electric vehicle chargers, and household batteries, identify local interruptions, route power around bottlenecks, and react to demand surges in milliseconds. But AI is the true revolution—the unseen force that gave rise to the smart grid. AI is more than just a helpful addition to the smart grid; it is its nervous system and brain, giving the whole energy infrastructure the ability to anticipate, learn, and adapt.

At its core, AI transforms the grid from a reactive entity into a predictive, decision-making engine. It digests and analyzes enormous amounts of data from asset performance, market signals, consumption trends, and weather forecasts to precisely plan electricity flows. It can balance loads across dispersed systems, predict demand surges before they happen, and release stored energy at the ideal time to prevent strain. Neural networks can identify abnormalities in transformer health before mechanical failures lead to blackouts; machine learning models trained on years of solar radiation data can forecast solar output hourly; and reinforcement learning agents can test thousands of dispatch strategies to determine the best course for cost, resilience, and emissions. The AI-powered grid is not just anticipatory, but also proactive.

This intelligence enables a radically new form of energy governance. We are shifting toward systems that are co-managed by utilities, prosumers, and algorithms rather than having centrally mandated management. Microgrids allow communities to function alone or dynamically connect to bigger systems. AI systems are capable of coordinating demand response from millions of smart devices, managing peer-to-peer energy exchanges, and optimizing collective storage dispatch. The end result is a distributed mesh grid that functions as a living bridge between intelligence and infrastructure rather than a rigid backbone. Energy justice is also made possible by this change since AI helps integrate distributed resources into larger systems by making them manageable. This means that community energy collectives, urban battery banks, and rural solar farms are not positioned at the perimeter of power but rather at its center (See Fig. 5.1).

However, this change is intellectual in nature rather than merely technological. The AI-powered smart grid calls into question the concept of infrastructure itself. It transforms from a physical object to a learning system that is influenced by data just as much as by steel and copper. It continuously adjusts its operational logic,

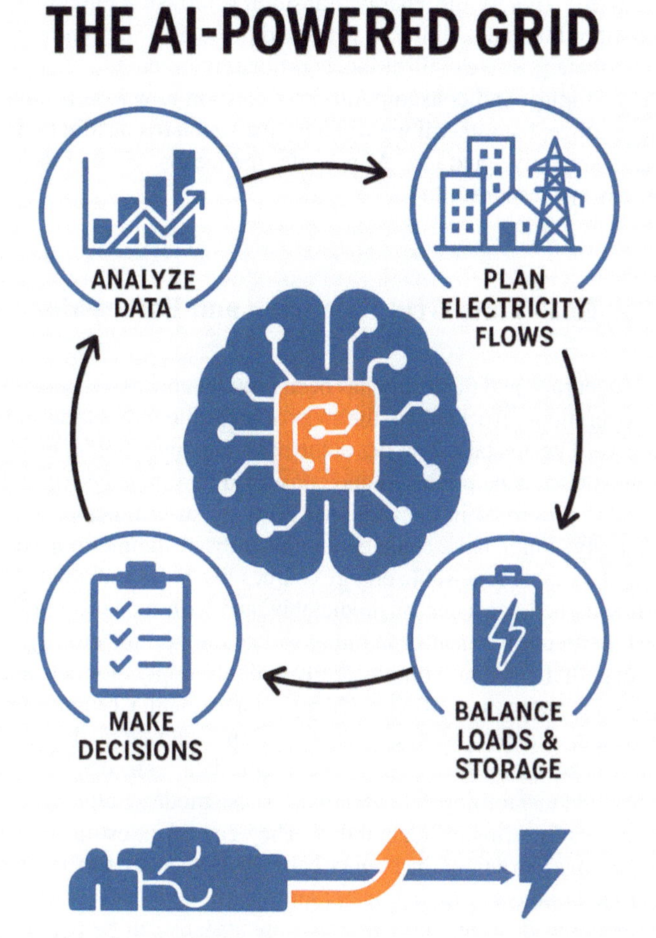

Fig. 5.1 AI transforms the power grid into a predictive, decision-making engine, analyzing data, planning flows, balancing loads, and optimizing storage to create resilient, efficient, and equitable energy systems for communities and utilities alike

learns from its mistakes, and adjusts to user behaviour. By doing this, it reinterprets resilience as agility rather than robustness, as anticipation rather than redundancy. Eliminating uncertainty is not the goal of the intelligent grid. It harnesses it, dances with it, and creates a new sort of dependability that comes from cooperation between technology, people, and the cycles of nature rather than from control.

There has never been a more pressing need for an adaptive, AI-enabled grid than now, as we stand on the brink of deep decarbonization, with entire economies promising to achieve net-zero and climate shocks becoming more common (Cheshmehzangi, 2025). This is the energy transition's infrastructure, which includes insights and algorithms in addition to wires and poles. It is the central neurological system of a climate-positive civilization, coordinating in real time the interaction of the sun, wind, storage, and consumption. The question now is not if the intelligent grid will materialize, but rather if we will have the vision to construct it in a way that is inclusive, regenerative, and just. AI provides the means. Writing the code for our shared future is our responsibility.

5.2 AI for Real-Time Grid Balancing and Forecasting

A new era of accuracy and flexibility has been brought about by the incorporation of AI into grid operations, especially in the areas of real-time forecasting and balancing. A straightforward yet unstable fact lies at the heart of contemporary energy systems: electricity must always be produced and consumed in almost perfect balance. This equilibrium was preserved in conventional grids by strict timetables and cautious planning. The dynamics have changed significantly in the modern era due to the introduction of variable renewable energy sources like solar and wind. This delicate balance is maintained consistently, predictably, and with the least amount of waste thanks to AI, particularly machine learning and neural optimization systems.

Predicting both the demand for electricity and the supply of renewable resources is one of the most potent uses of AI in energy management. Current energy demand projections are extremely accurate, both short- and long-term, thanks to machine learning models based on historical consumption data, calendar trends, socioeconomic factors, and meteorological conditions. These models outperform their statistical predecessors in terms of adaptability. They can adapt estimates on the fly to account for minute variations in human behaviour, such as shifting consumption owing to public holidays or higher use of air conditioners during heatwaves. Overgeneration, underproduction, and expensive imbalances can be prevented in areas where energy use is erratic or influenced by seasonal extremes.

On the supply side, AI plays a similarly crucial role in forecasting renewable generation. Despite being plentiful and clean, solar and wind energy are inherently unpredictable and weather-dependent. AI models can estimate solar farm and wind turbine production with surprising accuracy when trained on real-time satellite imagery, irradiance data, wind speed patterns, and weather forecasts. Not only are these minute-by-minute or hourly forecasts instructive, but they also have operational implications. In order to preserve stability, they enable grid operators to predict when and where generation will dip or surge and to proactively modify other sources, storage systems, or demand-side controls. AI prediction models are assisting in transforming intermittent renewable energy sources into reliable grid assets by reducing uncertainty (See Fig. 5.2).

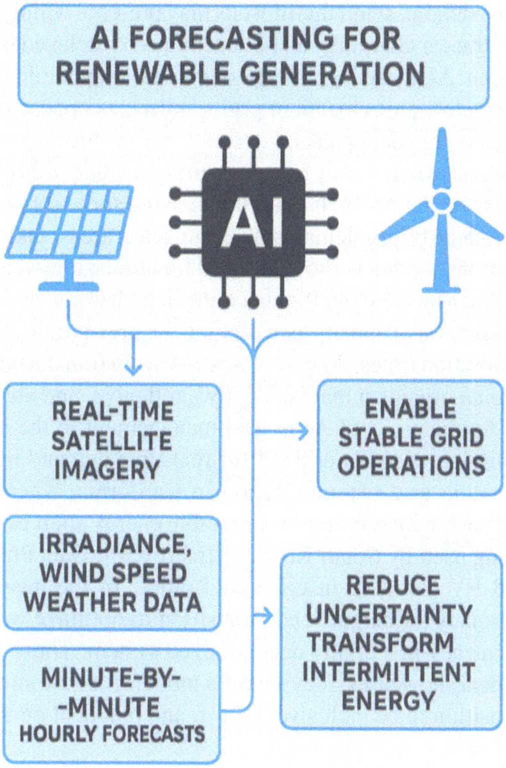

Fig. 5.2 AI forecasting transforms renewable generation reliability by analyzing satellite imagery, weather data, and wind patterns, enabling precise minute-by-minute predictions for stable grid operations and reducing uncertainty in intermittent energy supply

Balancing the grid in real time requires more than just good forecasts; it requires intelligent control systems that can respond swiftly and autonomously. One especially effective technique is reinforcement learning, a branch of machine learning in which algorithms discover the best course of action by making mistakes in simulated settings. By learning how to balance generation, storage, and load over intricate and dispersed grids, reinforcement learning agents can be trained to optimize dispatch techniques across thousands of situations. By incorporating real-time feedback from throughout the network, neural control systems go one step further and modify settings in response to shifting situations. These AI algorithms exceed conventional dispatch techniques by orders of magnitude in terms of speed and flexibility. They can schedule battery discharges, ramp gas peakers, manage distributed energy resources, and reroute power flows in milliseconds.

Dynamic load management, or changing consumption patterns to better match available supply, is one of the most revolutionary uses of AI (Wang et al., 2024). Distributed storage units, electric cars, and smart appliances can all be configured to react instantly to grid signals. These devices can be coordinated by AI platforms to improve grid dependability, smooth out consumption curves, and lower peak demand. For example, smart thermostats may pre-cool buildings prior to a supply decrease, and electric car charging can be accelerated or delayed based on projections of renewable output. AI allows us to shape demand as a dynamic variable, transforming consumers into active participants in grid stability, as opposed to constructing surplus capacity to meet transient peaks.

AI also plays a key role in ensuring resilience through self-healing networks in grid stress scenarios, such as abrupt demand surges or generator outages. These networks can identify problems, isolate impacted areas, and redirect electricity around interruptions—often without the need for human intervention—because they are equipped with sensors and intelligent controllers. During catastrophic weather events or cyber-attacks, this capacity is essential for preventing cascading outages and shortening restoration times. AI essentially assists in transforming the grid from a fragile system into an organism that is adaptive, reflexive, and able to defend itself.

Applications of AI in grid management in the real world highlight how revolutionary it is. AI is utilized for real-time demand response, fire risk prediction, and dynamic grid reconfiguration in California, where heat waves, wildfires, and the increasing integration of renewable energy sources present significant hurdles. AI is being used by South Korea's "Energy Big Data Platform" to manage solar, storage, and EV charging in congested cities. In the meantime, AI has been adopted by Germany's Energiewende transition to optimize wind energy dispatch and stabilize frequency in a highly decentralized system. These examples are not merely technological showcases; they serve as models for how intelligence might enable the energy transition to be inclusive, robust, and practical on a large scale (See Fig. 5.3).

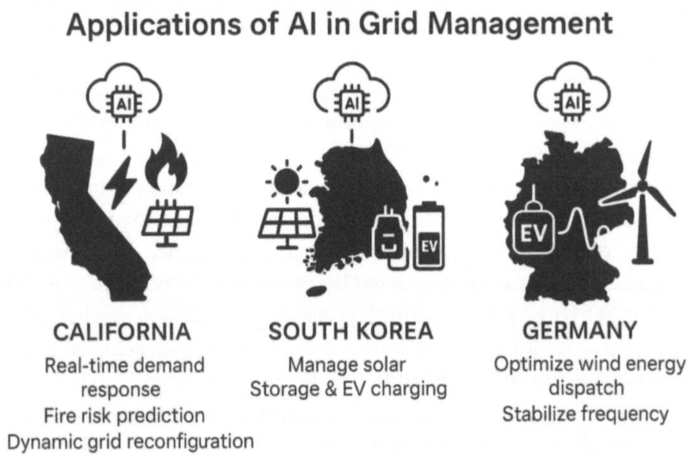

Fig. 5.3 AI is revolutionizing grid management worldwide—enabling California's demand response and fire risk prediction, South Korea's solar, storage, and EV optimization, and Germany's wind dispatch and frequency stabilization—building smarter, resilient, and inclusive energy systems

AI is a requirement in this new paradigm, not a luxury. With their decentralization, digitization, and decarbonization, current grids are too complicated to be handled solely by old software and human judgment. The only practical brain that can combine the enormous amounts of erratic input from the energy environment into activity that is responsive, logical, and forward-looking is AI. By using learning, prediction, and accuracy rather than force, it makes sure that every watt is in the correct place at the right time. This marks the beginning of an intelligent energy future in which balance is constantly found rather than preserved.

5.3 Distributed Intelligence: AI and the Integration of Decentralized Energy Resources

Energy decentralization is no more a theoretical trend; rather, it is a dynamic, expanding reality that is changing the way electricity is generated, distributed, and used. The energy architecture is changing from centralized monoliths to expansive networks of distributed energy resources (DERs) as rooftop solar panels grow more prevalent, house battery systems become more widespread, and communities construct microgrids and virtual power plants (VPPs). Energy sovereignty, resilience, and sustainability all stand to gain greatly from this transition, but it also adds an incredible amount of complication. Traditional grid systems cannot coordinate the actions of thousands, if not millions, of small, independently owned and operated assets in real time. Here, AI steps in as the organizing force of this distributed future, i.e., synchronizing, forecasting, trading, and optimizing at a scale and speed that human operators cannot achieve.

The emergence of the prosumer—people and companies that produce and store energy in addition to consuming it—is at the center of this shift. Their contributions, whether in the form of electric cars, solar panels, or batteries installed behind the meter, add variability to the grid in terms of both demand and generation. AI makes it possible for these inputs to be seamlessly coordinated, maximizing bi-directional energy flows based on weather forecasts, market signals, local consumption trends, and real-time supply. AI views DERs as orchestrated instruments in a dynamic energy symphony rather than as background noise, modifying their operation to optimize grid efficiency and financial return.

In this context, enabling peer-to-peer (P2P) energy trading is a particularly innovative use of AI. These decentralized networks, which are powered by AI algorithms and enabled by blockchain, let energy consumers trade electricity directly with one another. While AI platforms control pricing, match offers with demand, and make sure trades adhere to grid limits, homeowners with excess solar can sell it to neighbours instead of returning it to the utility at low feed-in tariffs. This increases economic inclusion and energy autonomy by promoting local energy economies where power remains in the community. These systems aren't theoretical. Projects in the Netherlands, Australia, and Japan have piloted AI-driven P2P markets with strong results, showing that such models can operate efficiently, equitably, and securely.

This level of coordinating granularity requires intelligence to reside near the grid's edge. In demand response and asset management, edge AI—i.e., AI embedded directly within local devices and infrastructure—is indispensable. AI-enabled smart meters, inverters, thermostats, and EV chargers may make minute judgments about when to charge, discharge, save, or use. Hyper-localized optimization is made possible and pressure on central systems is reduced as they react to situations in milliseconds. By sharing sensor data, implementing control techniques with low latency, and interacting with one another via the Internet of Things (IoT), these devices make up the modern grid's core system (Nematkhah et al., 2022; Pourroostaei Ardakani & Cheshmehzangi, 2023). In this setting, the grid becomes a collaborative, self-correcting ecosystem rather than a top-down structure.

The Tesla Autobidder technology, which is used throughout Australia's expanding fleet of virtual power plants, is a potent example of this paradigm in operation (Jones et al., 2021). Autobidder automatically bids stored energy from dispersed Tesla Powerwalls into wholesale electricity markets using machine learning (Desfosses, 2018). Hundreds or thousands of household batteries are combined into a single VPP, and their dispatch is optimized according to weather, grid demands, and real-time pricing signals. In addition to making money for users, this system offers vital grid functions like peak shaving and frequency management that are typically provided by large-scale generators (Huang et al., 2025). Distributed intelligence may compete with centralized infrastructure in terms of responsiveness and dependability, as the South Australian VPP, which is driven by Tesla's AI stack, has already shown.

Similar concepts are gaining traction in Japan and Europe. AI-enabled community grids in Germany are controlling residential energy sharing and local storage systems, assisting communities in becoming energy independent and lowering their dependency on fossil fuel plants. AI-powered microgrids in Japan, where seismic risks make energy resilience crucial, integrate solar, battery, and diesel systems to guarantee that neighbourhoods can function on their own in an emergency (Almihat & Munda, 2025). Over time, these systems learn and adjust, optimizing not only for cost and efficiency but also for supply continuity and carbon reduction.

Finally, using AI to integrate DERs changes the logic of energy itself rather than just resolving a practical issue. Communities no longer have to wait for central planning or top-down investment to become climate resilient. Energy becomes co-managed, democratized, and modular with distributed intelligence. AI transforms decentralization's complexity into an opportunity for hyper-local innovation, shared ownership, and energy systems that are more adaptable, inclusive, and able to handle the problems of the climate age. Not merely a new grid appears, but a new energy social contract where intelligence is dispersed as far as the resources it manages.

5.4 Decarbonization Through Intelligence: Policy, Access, and the Road Ahead

AI has become a key driver of the energy transition as the worldwide drive for decarbonization heats up (Lakhouit, 2025). In addition to increasing efficiency, its ability to predict, optimize, and coordinate across intricate, dispersed systems is paving the way for a completely new energy paradigm. Deep integration of renewable energy sources, the stabilization of unstable supply and demand, and the conversion of static infrastructure into dynamic, learning systems are all made possible by AI. As a result, it facilitates the quick adoption of clean energy technology, the stabilization of low-carbon networks, and the structural changes needed to achieve net-zero goals. Therefore, the intelligent grid is the glue that keeps the transition together and is not a supplementary instrument in the decarbonization strategy (See Fig. 5.4).

Yet, this transformation does not come without challenges. The effectiveness of AI depends on the quality, availability, and granularity of data. The capacity to deploy intelligent grid solutions fairly is hampered by the fragmented or non-existent data infrastructure in many places, especially in the Global South and underprivileged urban areas (Imran et al., 2024). Concerns about cybersecurity are also quite real. The grid is increasingly vulnerable to cyberattacks as it is increasingly networked and algorithmically controlled. Protecting vital infrastructure requires making sure AI systems are strong and resilient without sacrificing transparency and interoperability. Furthermore, legal frameworks frequently fall behind the rapid advancement of technology. Numerous energy markets are still set up according to antiquated regulations that fail to acknowledge the benefits of AI-led demand response, dynamic

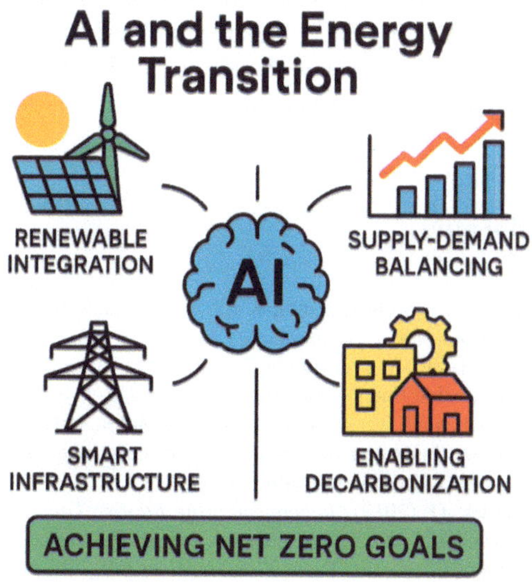

Fig. 5.4 AI is transforming energy systems by integrating renewables, balancing supply and demand, enabling smart infrastructure, and accelerating decarbonization—acting as the core driver (or the agent) of a resilient, net-zero future

pricing, and distributed generation. The entire potential of intelligent systems will continue to be underutilized in the absence of new policies.

It is also necessary to address more profound ethical and social issues. We must consider who gains when AI optimizes resource allocation, market involvement, and energy flows (Allam & Cheshmehzangi, 2024). Without careful planning, the smart grid might exacerbate already-existing disparities by giving those with money for solar panels and batteries cost savings and resilience while leaving others vulnerable to fluctuations. When algorithms are used in opaque ways or are trained on skewed data sets, they may unintentionally reflect or magnify systemic biases. To prevent a two-tiered energy future in which intelligence is treated as a privilege rather than a right, it is imperative that AI development be transparent, inclusive, and subject to public monitoring.

Open data and collaborative AI platforms play a critical role in combating this. We can democratize energy intelligence tools by granting open access to models, algorithms, and energy data. Communities can engage in local energy markets, construct their own microgrids, and allocate resources according to common goals rather than just financial gain. AI can facilitate leapfrogging in low-income and off-grid areas where the central grid may never reach by facilitating low-cost, high-efficiency electricity, organizing battery swaps, and optimizing solar microgrids. For these technologies to be reliable, accessible, and in line with human development objectives, collaborations between governments, academics, and civil society will be essential.

Lastly, we can argue that the smart grid driven by AI is more than just a technological advancement. It is a rethinking of how we produce, distribute, and control energy. By creating smarter, more connected, and human-centered systems instead of larger ones, it enables us to overcome our reliance on fossil fuels. This is the cornerstone of a sustainable energy future that is resilient, inclusive, and just in addition to being decarbonized. If we choose to construct with equity in mind, intelligence can be the enabler of equity rather than the antithesis of it. Adopting AI in the grid is only one of the challenges that lie ahead; another is integrating it with the principles and ideals required to drive a better future.

References

Allam, Z., & Cheshmehzangi, A. (2024). Technological Innovation and Sustainable Transitions. In *Sustainable Futures and Green New Deals* (pp. 55–79). Springer Nature Switzerland.

Almihat, M. G. M., & Munda, J. L. (2025). The Role of Smart Grid Technologies in Urban and Sustainable Energy Planning. *Energies, 18*(7), 1618.

Cheshmehzangi, A. (2025). Bridging Innovation and Sustainability: Global Insights on ICT Applications. In *Urban Innovation and Sustainability: ICT-based Interventions in Hainan, China* (pp. 23–43). Springer Nature Singapore.

Desfosses, D. (2018). *Jumping into the Future: The Transition Towards a Decentralised Energy System in Australia*. https://www.ie-ei.eu/Ressources/FCK/image/Theses/Energy%20System%20in%20Australia_David%20Desfosses.pdf

Gilrein, E. J., Carvalhaes, T. M., Markolf, S. A., Chester, M. V., Allenby, B. R., & Garcia, M. (2021). Concepts and practices for transforming infrastructure from rigid to adaptable. *Sustainable and Resilient Infrastructure, 6*(3–4), 213–234.

Huang, J., Li, H., & Zhang, Z. (2025). Review of Virtual Power Plant Response Capability Assessment and Optimization Dispatch. *Technologies, 13*(6), 216.

Imran, M. A., Zennaro, M., Popoola, O. R., Chiaraviglio, L., Zhang, H., Manzoni, P., van de Beek, J., Stewart, R., Cox, M. A., Mendes, L. L., & Pietrosemoli, E. (2024). Exploring the boundaries of connected systems: Communications for hard-to-reach areas and extreme conditions. *Proceedings of the IEEE, 112*(7), 912–945.

Jones, K. B., Franco, M., Mashke, K., & Pardee, S. A. (2021). How virtual power plants can advance electrification and mitigate infrastructure needs as we race to meet our climate challenges. *San Diego J. Climate & Energy L., 13*, 135.

Lakhouit, A. (2025). Investigating the hydrogen renaissance in the global energy transition with AI integration. *Energy Conversion and Management: X*, 101010.

Nematkhah, F., Aminifar, F., Shahidehpour, M., & Mokhtari, S. (2022). Evolution in computing paradigms for internet of things-enabled smart grid applications: Their contributions to power systems. *IEEE Systems, Man, and Cybernetics Magazine, 8*(3), 8–20.

Pourroostaei Ardakani, S., & Cheshmehzangi, A. (2023). The role of big data analytics in urban systems: review and prospect for smart transport and healthcare systems. *Big Data Analytics for Smart Transport and Healthcare Systems*, 1–26.

Wang, X., Wang, H., Bhandari, B., & Cheng, L. (2024). AI-empowered methods for smart energy consumption: A review of load forecasting, anomaly detection and demand response. *International Journal of Precision Engineering and Manufacturing-Green Technology, 11*(3), 963–993.

Chapter 6
Invisible Savings: AI for Energy Efficiency Across Built, Industrial, and Mobility Systems

6.1 Energy Efficiency as a Climate Imperative in the AI Era

The adoption of renewable energy, carbon capture technology, and electrification of industries and transportation are some of the ways that the goal of global decarbonization and net-zero emissions is frequently presented. However, one of the most practical, affordable, and scalable levers still accessible is energy efficiency, which is essentially using less energy to accomplish the same or better results. The International Energy Agency estimates that efficiency gains might provide more than 40% of the emissions reductions required to reach the targets set forth in the Paris Agreement (Höhne et al., 2021). Because efficiency directly lowers demand, lowers prices, and makes room for the slower ramp-up of clean energy generation, it becomes a key climate strategy rather than a secondary "nice-to-have".

However, implementation gaps have always plagued energy efficiency. Due to systemic and behavioural barriers, it has frequently remained an underutilized resource, even in cases where it is technically possible and economically advantageous (Battilani, 2014). Disjointed systems, each with its own controls and performance data, are used to govern built environments. Production, maintenance, and energy management departments in industrial facilities may operate in silos and hardly ever interact in real time. Fleet optimization in mobility systems is frequently constrained by antiquated routing software or human judgment calls (Tan et al., 2022). Because of this fragmentation, coupled systems seldom reach their full efficiency potential.

Historically, data, when it is accessible, has come too late to guide significant action. Monthly energy bills, periodic audits, and manual meter readings provide a post-mortem rather than a preventative measure. Energy is already spent by the time inefficiencies are discovered. The issue is made worse by manual controls and reactive decision-making, especially in complex or large-scale activities when there are more variables than a human can keep track of (Safari et al., 2024). Behavioural inertia is layered on top of these structural barriers: even when the economics obviously

© The Author(s), under exclusive license to Springer Nature Singapore Pte Ltd. 2025
A. Cheshmehzangi, *AI Agency in Climate Emergency*, Urban Sustainability,
https://doi.org/10.1007/978-981-95-4192-8_6

support efficiency, decision-makers continue with "good enough" operations because the effort required to change appears to outweigh the perceived gain.

These obstacles are being quickly removed by the AI era, which makes it possible to go from trailing indicators to real-time, predictive optimization. Intelligent automation can balance conflicting goals, modify system parameters, and continuously monitor conditions—often without requiring human intervention. In order to optimize heating, cooling, lighting, and ventilation in buildings on the go, AI-driven energy management platforms increasingly incorporate sensor data, weather forecasts, occupancy patterns, and grid signals (Vinoth & Shibi, 2025). Predictive algorithms can detect process bottlenecks, estimate peak load times, and adjust equipment performance in industrial settings before inefficiencies worsen. AI-powered dynamic routing in mobility systems minimizes fuel use, eases traffic and can work with electric fleet charging infrastructure.

In this situation, AI's ability to combine precision control, pattern recognition, and prediction makes it a strategic lever (Ramya et al., 2024). Systems can react to changing conditions by using prediction, such as pre-cooling a building before a heatwave or modifying industrial operations before a spike in demand. Pattern recognition reveals inefficiencies that are not evident to human study, such as the subtle relationships between energy intensity and machine vibration or between citywide patterns of fuel usage and traffic movement. By ensuring that interventions are precisely calibrated rather than blunt-force, precision control optimizes for cost, carbon intensity, comfort, and productivity all at once.

Importantly, when AI is used throughout industries rather than in discrete fields, its benefits for energy efficiency increase dramatically. There are connections between the built environment, industry, and mobility systems that, when combined, can result in significant energy savings. For instance, a smart district energy system can reduce peak generation requirements by dynamically adjusting thermal loads according to building occupancy and industrial process schedules (Cheshmehzangi & Pourroostaei Ardakani, 2024). By scheduling deliveries during off-peak energy times, AI-powered logistics solutions can prevent grid stress and save operating expenses for energy suppliers and transport operators alike. By coordinating urban mobility planning with public building operations, peak transportation demand can be shifted to times when renewable energy supplies are plentiful.

More than simply data sharing is needed for this cross-sector orchestration; a systems-thinking approach backed by open data standards, interoperable platforms, and governance frameworks that reward efficiency as a common goal is also necessary. AI serves as the glue, converting knowledge from one industry into useful optimization in another. To ensure comfort and energy savings, a building energy management system that "talks" to a city's traffic control network, for example, may pre-condition transit hubs depending on real-time passenger inflows. Overall consumption patterns could be smoothed by scheduling energy-intensive jobs during periods when demand for electric vehicle charging is low at industrial facilities connected to local grid and mobility systems.

The operational constraints that formerly limited energy efficiency are eroding, strengthening the climatic case for energy efficiency in the AI era (Pimenow et al.,

2024). We may transition from static, one-time retrofits to dynamic, ever-improving systems that change as circumstances do. Energy efficiency stops being a side project and instead becomes an integrated, flexible aspect of production, transportation, and infrastructure. By redefining efficiency as a strategic climate asset, it not only lowers emissions directly but also makes it possible for renewable energy sources to become more widely used by balancing supply and demand.

Scaling these intelligent efficiency solutions beyond pilot projects and showpiece buildings is now essential. This calls for spending money on the digital infrastructure, cross-sector collaborations, and governance structures that allow AI technologies to work as cohesive ecosystems in addition to the technology themselves. Rethinking metrics is also necessary; efficiency benefits should be quantified in terms of avoided emissions, enhanced resilience, and system-level optimization rather than just kilowatt-hours saved.

Our ability to extract more value from each unit of energy will become increasingly important as net-zero deadlines approach, determining the gap between goal and reality (Sharma & De, 2023). AI is a force multiplier in this race, not a magic wand. AI has the potential to transform the long-underutilized promise of energy efficiency into a key component of climate action by making predictions more correctly, acting faster, and coordinating more widely than human-managed systems ever could. Its invisible savings—across factories, transportation networks, and buildings—may end up being some of our most effective climate measures since they are ingrained in the systems we already utilize on a daily basis.

By adopting this change, energy efficiency may transform from a reactive, incremental practice to a proactive, systemic force for decarbonization, which will not only cut waste but also change the way our built, industrial, and mobility systems interact to support a future free from climate change. We have the tools from the AI era, and we must use them to meet the climate imperative (See Fig. 6.1).

6.2 Smarter Buildings: AI for Automated Energy Management

Buildings are among the largest consumers of energy globally, responsible for roughly 30–40% of total end-use energy and a comparable share of greenhouse gas emissions. Heating, ventilation, and air conditioning (HVAC) systems are the main attractions in many metropolitan settings. Lighting, appliances, and plug loads are next in line. Occupancy patterns—people's arrival and departure times, the level of activity in various areas, and the interplay between indoor comfort preferences and outdoor climate conditions—have a significant impact on how these loads change throughout the day. Historically, reactive maintenance, predetermined temperature objectives, and static schedules were used to address these variables, which led to inefficiency. This is altered by AI-powered energy management (Cheshmehzangi,

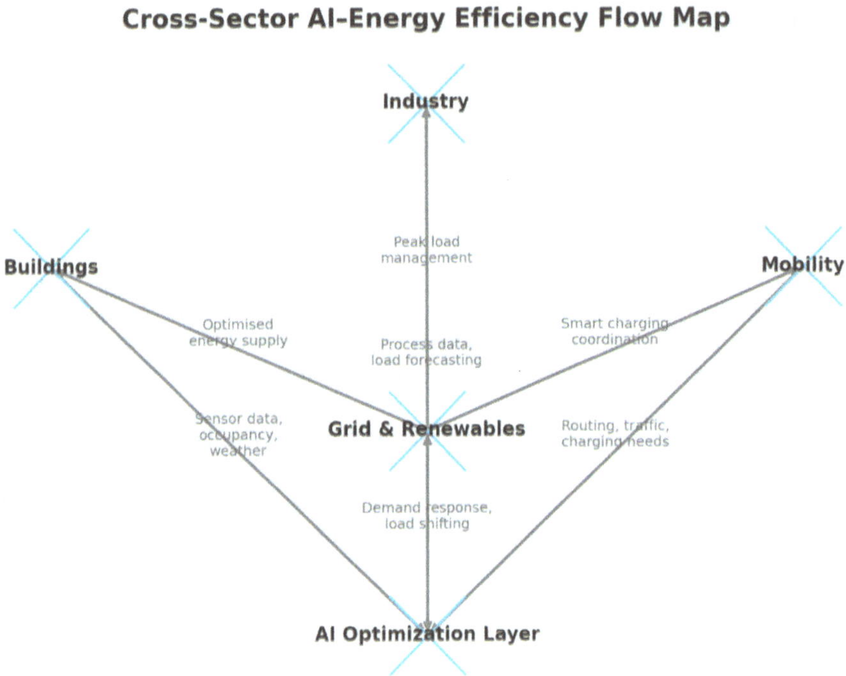

Fig. 6.1 Summary of Cross-Sector AI-Energy Efficiency Flow Map

2026), which learns a building's complex "energy signature" and uses it to coordinate systems for the least amount of waste and the most comfort.

The energy consumption profile of a building is a dynamic fingerprint that is influenced by its use (office, school, retail), the behaviours of its occupants, the local climate, and the system performance. While lighting and plug loads are more common in temperate regions or in buildings with a lot of natural light, HVAC loads usually predominate in locations with extreme temperatures. Building Energy Management Systems (BEMS) with AI capabilities turn this knowledge into a dynamic, living system. One of AI's most potent contributions is predictive control, which anticipates demand and modifies systems before peaks occur by utilizing real-time sensor inputs, weather forecasts, and previous energy data. This prolongs equipment life by preventing overcooling or overheating, lowering startup surges, and facilitating seamless load changes.

The ability of AI to recognize patterns is also advantageous for thermal comfort modelling. AI can predict how different areas of a building react to temperature, humidity, and occupancy, then adjust airflows, setpoints, and schedules accordingly, eliminating the need for the universal thresholds seen in traditional HVAC controls. Similar to this, adaptive lighting systems can change the colour temperature and brightness according to occupancy, the time of day, and even the kind of activity being done, in addition to daylight levels. This can be used in the workplace to create calmer environments later in the day and increase productivity during peak hours by adjusting light intensity to circadian rhythms.

6.2.1 The Sensor-Driven Intelligence Layer

A vast flow of real-time data is essential to the intelligence of contemporary AI energy management. The Internet of Things (IoT) and sensors can help with this. AI algorithms that are constantly learning the building's behaviours are fed data via networks of temperature, humidity, CO_2, occupancy, and light sensors. AI can do simulations in real time and evaluate control strategies before implementing them on the physical system when combined with digital twins, which are virtual representations of the building and its systems. This enables quick optimization and iteration without disturbing residents.

Long-term performance implications can also be modelled with digital twins. AI, for example, may model how altering an HVAC schedule during the summer impacts thermal comfort, maintenance requirements, and the equilibrium with other systems, such dehumidifiers, in addition to the immediate energy consumption. Facility managers can steer clear of unforeseen trade-offs, like cutting energy in one place while raising it in another, by adopting this systemic perspective.

6.2.2 Aligning Buildings with Renewable Energy

Synchronizing building loads with renewable generation and storage is one of AI's most revolutionary contributions to energy management. Buildings with AI-driven control can react instantly to changes in solar and wind power, which are inherently volatile. AI is able to move non-essential loads to periods when renewable output is strong, and storage is available by integrating with on-site photovoltaic (PV) panels, battery storage systems, and even electric vehicle (EV) charging stations.

When solar output peaks in the middle of the day, AI may pre-cool a commercial tower with rooftop photovoltaics, for instance, lowering the need for grid electricity in the late afternoon when demand soars. In a similar vein, AI can schedule EV charging with excess solar power, guaranteeing that cars are charged when clean energy is plentiful rather than when the grid is stressed. In addition to cutting emissions, this strategy can lower operating costs by utilizing time-of-use pricing.

AI-driven BEMS has been used in a number of high-performance commercial towers in Singapore, including the Marina One complex, to reduce cooling energy consumption by as much as 30%. By combining occupancy-based controls, thermal storage, and chiller plant optimization, these systems constantly improve operation by learning from the building's performance. Because of the city's high humidity and density, HVAC optimization is particularly important. AI has proved essential in striking a balance between efficiency and comfort without the need for manual micromanagement.

Large office campuses and mixed-use developments around Europe have implemented AI-driven energy management to handle not only lighting and HVAC systems but also microgrids and district heating. AI is used, for example, to predict energy consumption and renewable generation in a smart office development in Amsterdam, allowing for coordinated load shifting across several buildings. To maximize financial gains and maintain grid stability, the AI system also interacts with the local energy market, selling stored electricity at times of high price and purchasing it during times of low price.

University campuses have emerged as living laboratories for net-zero building operation. In a number of instances, campuses have been able to create microgrids that function primarily independently of the main grid during the day thanks to AI-driven control that allows them to balance solar PV production, battery storage, and heating/cooling plants in real time. In order to fully integrate the building into the larger energy ecosystem, these systems also have demand-response capabilities, which automatically lower or shift loads when grid operators detect stress events.

6.2.3 Toward Autonomous Energy Buildings

AI's trajectory in building energy management is expected to be completely autonomous, with only strategic decision-making and supervision requiring human involvement. The operation of these "self-driving" buildings would be continuously improved by internal performance data as well as external signals from the grid, the weather, and market conditions. Manual control gives way to policy-setting, where humans establish financial priorities, sustainability goals, and comfort thresholds that the AI system then optimizes toward.

The shift to climate-positive buildings—structures that generate more clean energy than they need annually—is also supported by this progression. Such results are now possible in both new and renovated buildings thanks to AI that coordinates creation, storage, and demand in almost real-time. The building is converted from a passive energy consumer into an active, responsive node in the urban energy network by combining predictive control, real-time optimization, and renewable integration (See Fig. 6.2).

Scaling these capabilities from upscale developments and experimental projects to the larger building stock, which includes smaller and older structures, is currently the difficulty. The possibility of retrofitting AI-driven energy management into existing buildings will increase as sensor costs decrease, cloud-based AI platforms develop, and interoperability standards advance. In addition to enhancing comfort, resilience, and economic performance for both owners and tenants, this might lead to significant efficiency gains throughout cities, assisting in the reduction of the gap to net-zero targets.

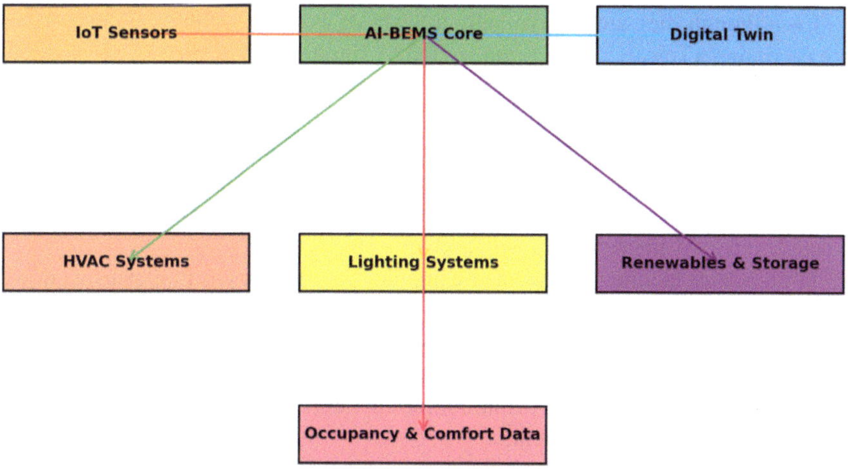

Fig. 6.2 A model for AI-Driven Building Energy Management System Integration

6.3 Industrial Intelligence: AI for Process Efficiency and Predictive Maintenance

Despite being the backbone of the global economy, energy-intensive industries including manufacturing, mining, data centers, and heavy industry are also among the biggest emitters of greenhouse gases. Large data centers use gigawatt-hours of electricity a year only to keep servers operating and cooled, and the manufacture of steel, cement, and chemicals collectively contribute significantly to the world's CO_2 emissions. In the past, a combination of rule-based automation, manual supervision, and recurring equipment inspections have been used for operational control in these sectors. Although these methods maintain production, they are reactive and frequently ineffective, rarely optimizing systems in real time and only addressing issues after they arise. AI changes this by making it possible to control industrial processes in a predictive, adaptable, and integrated manner.

The heating, cooling, and material processing processes in heavy industries use a lot of energy. For instance, kilns used in the production of cement run at temperatures beyond 1,400 °C, necessitating careful thermal control to prevent fuel waste or lowered product quality. Chemical plants manage intricate, multi-stage reactions that can be energy-hungry if not closely controlled, while blast furnaces and rolling mills in steel plants must strike a balance between throughput and temperature consistency. The complex patterns of heat transfer, fluid movement, and material behaviour in these systems are learned via AI-driven process optimization. AI may modify control variables, including burner rates, airflow, and cooling cycles, minute by minute to maintain ideal conditions with the least amount of energy waste by combining sensor readings, historical performance data, and predictive models.

Even though they are not typically thought of as "heavy industry," data centers are becoming a key area of AI optimization. A significant portion of their electricity consumption is for cooling. AI systems can reduce energy consumption without running the danger of overheating by dynamically regulating chilled-water systems, fan speeds, and airflow depending on real-time server demand and environmental conditions. The notable use of reinforcement learning by Google's DeepMind in its data center cooling systems resulted in a 40% cooling energy savings and established a standard for AI-managed infrastructure efficiency.

In industrial facilities, unplanned equipment failures result in expensive downtime, energy waste, and occasionally safety risks. AI-driven anomaly detection drives predictive maintenance, which replaces reactive scheduling with a condition-based strategy. From conveyor motors in mining operations to compressors in petrochemical plants, sensor arrays assess vibration, temperature, pressure, acoustic signatures, and electrical load on equipment.

These data streams are analyzed by machine learning algorithms to provide a "normal" operational baseline for every asset. AI systems can detect anomalies that deviate from this baseline, including minute temperature spikes or mild increases in vibration frequency, before they become serious enough to cause a failure. This prevents production halts and lowers the additional energy consumption that frequently follows mechanical breakdown by enabling maintenance teams to arrange interventions during scheduled downtime. Predictive maintenance lowers the environmental and logistical costs of unscheduled repair trips for mining operations, when equipment operates in remote, hostile areas. It stops cascading defects in production, where a single machine failure lowers the efficiency of several lines until repairs are made.

Production timetables, batch processing cycles, and the requirement to synchronize numerous systems are common constraints on industrial processes. For example, in mining, crushing, and milling activities must be coordinated with transport and stockpiling schedules, while in chemical manufacture, certain processes must adhere to rigorous timing sequences to guarantee product uniformity. AI can optimize these schedules to balance workloads, eliminate bottlenecks, and move energy-intensive tasks to periods with cheaper electricity or more renewable energy. In industries like food manufacturing, where refrigeration demand may be adjusted and synchronized with other plant processes, scheduling optimization is particularly beneficial. AI may evaluate "what if" possibilities, such as altering operating conditions, batch sizes, or step sequencing, by modelling production flows using digital twins. This allows AI to identify configurations that produce the same output with less downtime and energy.

An excellent illustration of AI used to improve industrial process efficiency is Siemens' MindSphere platform. It combines cloud analytics, AI algorithms, and Internet of Things sensor networks to continuously monitor and optimize industrial assets. MindSphere has made it possible for industrial plant operators to identify production line efficiency losses, instantly modify heating and cooling profiles, and

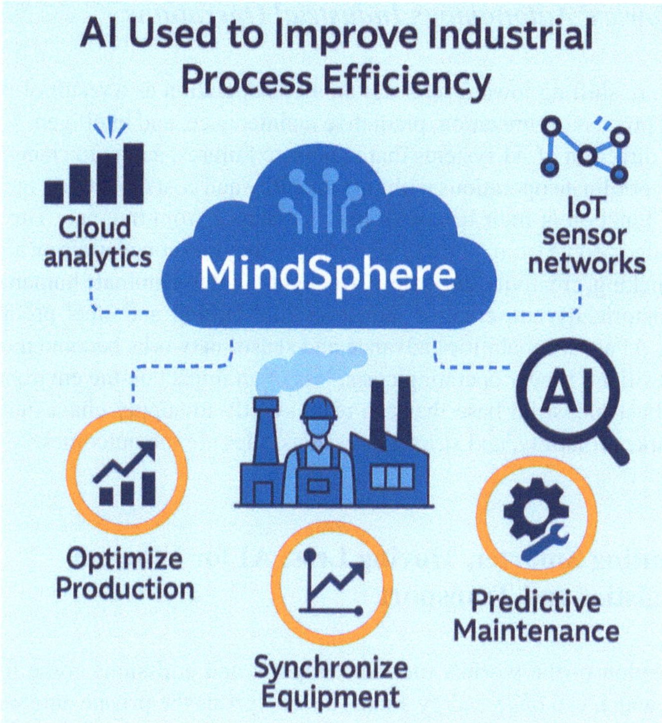

Fig. 6.3 Siemens' MindSphere integrates cloud analytics, AI, and IoT sensor networks to optimize production, synchronize equipment, and enable predictive maintenance, boosting efficiency, reducing downtime, and extending the lifespan of critical industrial assets

synchronize equipment to cut down on idle time. By automating maintenance schedules and identifying wear trends early, its predictive analytics modules also increase the lifespan of vital gear (See Fig. 6.3).

The DeepMind accomplishment in data center cooling by Google continues to serve as a benchmark for AI in infrastructure optimization. The AI discovered intricate cause-and-effect links between environmental factors and cooling system performance through the application of reinforcement learning. In addition to little improvements, the outcome was a significant move toward self-optimizing infrastructure, in which the system continuously learns and gets better without direct human assistance. Large-scale manufacturing complexes and industrial refrigeration warehouses are two examples of high-energy-load facilities that are now investigating this paradigm.

6.3.1 Toward Autonomous Industrial Operations

Industries are shifting toward semi-autonomous operation as a result of the confluence of AI process optimization, predictive maintenance, and intelligent scheduling. Under the direction of AI systems that anticipate failures, adjust to changing conditions, and coordinate operations with sustainability and cost objectives, facilities will be able to function at their best with little assistance from humans. The goal is to release skilled people from routine monitoring so they can concentrate on higher-level decision-making, creativity, and safety oversight—not to eliminate human oversight.

Even historically conservative industries like mining and steel production are embracing AI as its capabilities advance and sensor networks become more affordable and resilient. Lower operating costs, less of an impact on the environment, and a more robust industrial base that can react swiftly to supply chain interruptions, energy market volatility, and stricter emissions rules are the outcomes.

6.4 Routing Smarter, Moving Less: AI for Efficient Logistics and Transport

A large portion of the world's fuel consumption and emissions come from transportation, which is a huge energy sink. This comprises the private automobiles that rule the streets of cities, the vast logistical fleets that transport commodities across continents, and the public transportation networks that transport millions of people every day. Motion—much of it superfluous, inefficient, or ill-timed—is the common denominator. AI-driven optimization reinterprets the problem as one of moving with intelligence, accuracy, and purpose rather than just moving less.

One of the main tools for reducing energy use in logistics is route optimization. AI systems create routes that minimize distance, idle time, and stop-start inefficiency by ingesting traffic data, weather forecasts, road limitations, and delivery priority. This is furthered by fleet scheduling algorithms, which maintain trucks and vans running close to their maximum capacity by balancing driver hours, vehicle load factors, and maintenance periods. Routes and schedules are continuously adjusted as conditions change, preventing traffic jams, unplanned closures, and energy-consuming detours. This is not a one-time computation.

Modal shift methods, in which passengers or freight are dynamically directed to the most energy-efficient mode available, are also made possible by AI. When road networks are overloaded, this may entail directing commuters to shared shuttles, transit, or micromobility options, or it may entail moving cargo from trucks to rail for long-haul sections. By coordinating these decisions over thousands of separate voyages, the system is intelligent enough to generate cumulative gains that would be impossible for any one player to accomplish on their own. These advantages are increased via predictive traffic modelling. AI predicts congestion hotspots hours or even days ahead of time using historical and real-time data streams. In order to minimize fuel consumption and the possibility of missing delivery windows, navigation systems can then suggest departure times and routes that steer clear of these slowdowns. Predictive demand models enable public transportation operators to proactively modify vehicle allocation and service frequency, aligning supply with anticipated passenger loads.

In urban logistics, machine learning also drives freight consolidation, which groups the products of several suppliers into fewer, more compact vehicles. With the help of AI-driven demand forecasting, this strategy can significantly cut down on the number of journeys into crowded city centers. When paired with cargo bikes for last-mile delivery and micro-distribution hubs, it creates a cleaner and faster hybrid logistics network. Another breakthrough made possible by AI is demand-responsive public transportation (Pourroostaei Ardakani & Cheshmehzangi, 2023). Micromobility solutions such as e-bikes and scooters can be smoothly integrated with shuttle services and autonomous minibuses that can be deployed dynamically based on passenger requests (Krichi et al., 2025). This adaptability ensures accessibility and convenience while removing the waste of operating underutilized fixed routes, especially in low-density areas or during off-peak hours.

Lastly, being efficient in logistics and transportation means being smarter, not less. AI enables systemic energy reductions that cut across industries and scales by giving movement a purpose, synchronization, and context awareness. A mobility ecology that promotes economic vitality without entangling us in energy waste results when commodities arrive on schedule and without excessive mileage, buses arrive when and where they are needed, and personal travel adjusts fluidly to the network's pulse. A world where every kilometer travelled matters and none is wasted is what smarter routing and reduced movement promise.

References

Battilani, A. (2014, June). Limited access to resources: Challenges or opportunities?. In *XIII International Symposium on Processing Tomato 1081* (pp. 27–40).

Cheshmehzangi, A. (2026). *Generative AI-Powered Urban Digital Twins: Pioneering Environmental Solutions for Sustainable Intelligent Cities*. Springer.

Cheshmehzangi, A., & Ardakani, S. P. (2024). Embracing the digital twin paradigm for urban sustainability. In *Digital Twin Computing for Urban Intelligence* (pp. 1–11). Springer Nature Singapore.

Höhne, N., Kuramochi, T., Warnecke, C., Röser, F., Fekete, H., Hagemann, M., Day, T., Tewari, R., Kurdziel, M., Sterl, S., & Gonzales, S. (2021, December). The Paris Agreement: Resolving the inconsistency between global goals and national contributions. In *Climate policy after the 2015 Paris climate conference* (pp. 30–46). Routledge.

Krichi, M., Noman, A. M., Fannakh, M., Raffak, T., & Haidar, Z. A. (2025). Sustainable electric micromobility through integrated power electronic systems and control strategies. *Energies, 18*(8), 2143.

Pimenow, S., Pimenowa, O., & Prus, P. (2024). Challenges of artificial intelligence development in the context of energy consumption and impact on climate change. *Energies, 17*(23), 5965.

Pourroostaei Ardakani, S., & Cheshmehzangi, A. (2023). The role of big data analytics in urban systems: review and prospect for smart transport and healthcare systems. *Big Data Analytics for Smart Transport and Healthcare Systems*, 1–26. Springer.

Ramya, J., Yerraguravagari, S. S., Gaikwad, S., & Gupta, R. K. (2024). AI and Machine Learning in Predictive Analytics: Revolutionizing Business Strategies through Big Data Insights. *Library of Progress-Library Science, Information Technology & Computer, 44*(3).

Safari, A., Daneshvar, M., & Anvari-Moghaddam, A. (2024). Energy intelligence: A systematic review of artificial intelligence for energy management. *Applied Sciences, 14*(23), 11112.

Sharma, N., & De, P. K. (2023). *Towards Net-Zero Targets: Usage of Data Science for Long-Term Sustainability Pathways*. Springer.

Tan, W., Sun, Y., Ding, Z., & Lee, W. J. (2022). Fleet management and charging scheduling for shared mobility-on-demand system: A systematic review. *IEEE Open Access Journal of Power and Energy, 9*, 425–436.

Vinoth L. I., & Shibi, B. (2025). AI-Driven Smart Energy Management Systems for Optimized Renewable Energy Utilization in Urban Smart Grids. Preprint paper, https://doi.org/10.21203/rs.3.rs-6326251/v1

Chapter 7
Moving Clean: AI for Decarbonizing Transport and Electrifying Mobility Systems

7.1 Transport at a Tipping Point: The Urgency of Decarbonizing Mobility

Transportation is responsible for around 25% of global energy-related greenhouse gas emissions, making it one of the biggest single contributors to CO_2 emissions. This footprint is dominated by road vehicles, including the hundreds of millions of private automobiles on the road worldwide, long-haul freight, and huge logistical fleets. Even while public transportation is efficient per passenger kilometer, it frequently uses fossil fuels in many areas, and shipping and aviation add even more weight. Because mobility is ingrained in economies and daily life, the sector's emissions are stubbornly high; therefore, lowering its environmental impact calls for more than simply cleaner engines—it also calls for a reconsideration of how and why we travel.

The urgency is amplified by the pace of climate change and the slow turnover of vehicle fleets. Diesel freight trucks frequently live even longer than cars, which can be sold today and be on the road for more than ten years. This implies that high-emitting assets will remain dominant well into the 2030s unless swift action is taken. Incremental efficiency gains are insufficient to meet the Paris Agreement targets, which call for a dramatic reduction in transportation emissions within the next ten years (Bhattacharya et al., 2024). Yet, structural obstacles stand in the way of deep decarbonization. The main one is ingrained reliance on personal vehicles, especially in nations with low-density zoning, suburban expansion, and inadequate public transportation funding, which provide few practical alternatives to private automobiles. Fragmented governance amongst agencies can lead to inefficiencies, coverage gaps, and poorly integrated services that force people back into their automobiles, even in places with robust transport networks. On the freight side, vehicles frequently run empty on return journeys due to compartmentalized logistical activities, wasting capacity and fuel.

The absence of infrastructure for low-carbon vehicle charging and refuelling is another major obstacle. Although the use of electric vehicles (EVs) is increasing

© The Author(s), under exclusive license to Springer Nature Singapore Pte Ltd. 2025
A. Cheshmehzangi, *AI Agency in Climate Emergency*, Urban Sustainability,
https://doi.org/10.1007/978-981-95-4192-8_7

in various markets, range anxiety is still a problem in areas with few, sluggish, or incompatible chargers. Since battery technology and other fuels like hydrogen are still in their infancy, the problem is significantly more severe for heavy-duty freight, aircraft, and shipping. Electrification runs the risk of stagnating before it can realize its revolutionary promise in the absence of dependable, extensive infrastructure (See Fig. 7.1).

There is no denying the potential of electrification and modal change. Particularly when driven by clean grids, electrifying passenger automobiles, buses, and freight trucks can reduce exhaust emissions. Reclaiming urban space and lowering emissions and traffic can be achieved by switching to public transportation, walking, bicycling, and shared mobility. However, these fixes cannot be seen as straightforward replacements. Complex battery supply chains, grid upgrades to manage charging loads, and long-term planning for fair access are all part of electrification. In order to make sustainable solutions the norm, modal shift necessitates political will, cultural change, and urban environment redesign.

AI enters the picture as a necessary orchestrator of complexity rather than as a panacea. Through the integration of traffic projections, weather trends, and current

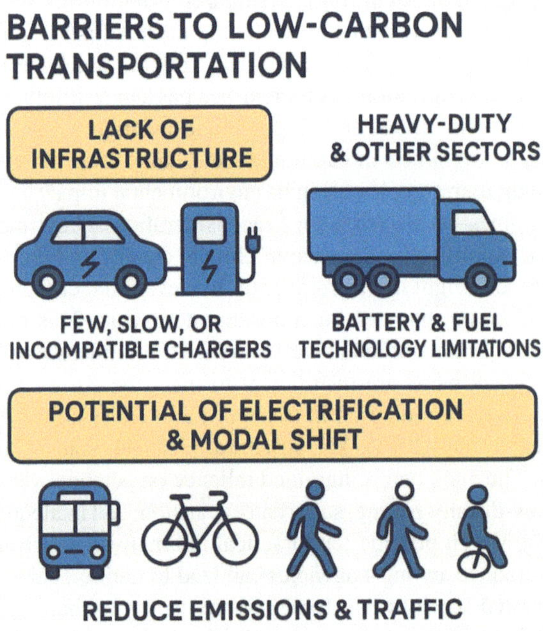

Fig. 7.1 Infrastructure gaps, technology limitations, and sector-specific challenges hinder low-carbon transport. Electrification and modal shift can cut emissions and congestion, but require political will, cultural change, and equitable access to charging and alternative mobility options

conditions, AI can optimize routes to minimize energy use for both individual vehicles and entire fleets. AI may be used to forecast infrastructure by predicting when and where charging demand will peak, allowing chargers to be strategically placed to correspond with adoption rates and travel trends. By using predictive analytics to match charging schedules with the production of renewable energy, electrification's environmental benefits may be completely realized. AI can also be used to manage modal transition tactics. Machine learning can model the effects of new transport routes, optimize service frequencies for maximum adoption, and analyze mobility data to discover underserved locations (Mittal & Ukkusuri, 2025). AI-driven logistics platforms in freight can minimize delays in rail-road-sea interface coordination, reduce empty runs, and combine goods among operators. In fragmented, human-managed systems, these applications generate efficiencies that are otherwise hard to measure.

It is crucial to characterize transportation decarbonization as a systemic change rather than just a technical replacement (Argyriou & Barry, 2021). Congestion, land use, and embedded emissions in manufacturing would not be addressed by switching to electric vehicles for all fossil fuel-powered vehicles. Similarly, if charging stations are installed without enhancing grid resilience, emissions may be transferred upstream to electrical generation that uses fossil fuels. Electrification, modal shift, supply chain decarbonization, and behavioural change are all components of true transformation.

The interdependencies of infrastructure, technology, policy, and human behaviour are acknowledged by this systemic perspective (Hellström, 2007). By synthesizing massive data streams, finding synergies, and facilitating coordinated decision-making among stakeholders, AI serves as the connecting thread. It can predict demand for the raw materials, energy flows, and maintenance cycles that support the whole mobility ecosystem, in addition to cars and chargers. By doing this, it lowers the possibility of assets becoming stranded, foresees bottlenecks, and promotes the fair and effective use of resources.

Business as usual locks in decades of emissions, but swift, well-managed change can result in safer streets, cleaner air, and robust urban economies. Transportation is at a tipping point. Redesigning mobility systems to benefit both people and the environment is more urgent than just replacing engines. AI provides the strategic intelligence needed to manage this shift, making the difficulty of decarbonization a strength rather than a weakness (Lah, 2017). By doing this, it makes possible a future in which mobility contributes to climate stability and prosperity.

7.2 AI in Public Transit Optimization and Shared Mobility

Decarbonizing urban transportation requires the use of public transportation and shared mobility, but their effectiveness rests on their capacity to balance supply and demand while preserving coverage and dependability. AI gives mobility providers and transit agencies the predictive and adaptive power they need to strike this balance, enabling them to make decisions in real time that enhance service quality, save operating costs, and increase accessibility. For instance, demand forecasts and route planning heavily rely on machine learning. AI can predict passenger volumes at particular stops and times by analyzing past ridership data in conjunction with outside variables like the weather, local events, economic activity, and seasonal trends. By modifying bus and train routes ahead of time, these projections enable agencies to reduce underutilized trips and increase service where demand will spike. Without the trial-and-error of conventional planning cycles, this predictive capability allows for improved resource allocation in big networks.

By continuously adjusting timetables and vehicle allocations based on current conditions, dynamic scheduling goes one step further. AI can automatically reroute other vehicles or modify frequency to prevent service interruptions in the event that traffic congestion delays a bus. To maintain constant headways and reduce crowding, fleet dispatch systems redeploy buses, trains, or trams where they are most required using real-time GPS, passenger counts, and control center inputs. This is especially helpful during outages, when quick fixes can prevent networks from experiencing undue delays.

AI is also used by shared mobility platforms, such as ridesharing services, on-demand shuttles, and fleets of e-bikes and e-scooters, to optimize matching, routing, and coverage. Algorithms maximize seat occupancy and minimize diversions by matching passengers with cars going in compatible directions. In order to prevent shortages in high-use areas and oversupply in low-demand zones, AI models for micromobility forecast where demand will rise and realign bikes or scooters accordingly. This guarantees improved user experience and increased utilization rates. By locating underserved locations where service gaps restrict mobility alternatives, AI can also address transit equity. AI shows "transit deserts" and directs targeted service expansions or last-mile solutions, such on-demand shuttles connected to rail hubs, by mapping demographic data, land use trends, and travel demand. By doing this, it contributes to ensuring that communities with low incomes, remote locations, and limited mobility do not lose out on the advantages of an effective transportation system (See Fig. 7.2).

AI-led optimization shows promise in a number of cities. Using AI to arrange multimodal trips and dynamically price choices to balance demand, Helsinki's Whim mobility-as-a-service platform combines public transportation, taxis, auto rentals, and micromobility into a single app. Using farecard data and passenger load sensors, Singapore's Land Transport Authority uses AI to modify bus frequencies in real time. Predictive analytics is used by Los Angeles Metro to optimize bus deployment, integrating real-time incidents, boarding patterns, and traffic data to increase network reliability. AI turns shared mobility and public transportation into flexible, demand-responsive systems by fusing adaptive control with predictive insights. In addition to increasing efficiency and lowering emissions, this also increases public trust, which motivates more people to select environmentally friendly transportation options over private vehicles. A more cohesive, just, and robust urban mobility ecology is the end consequence.

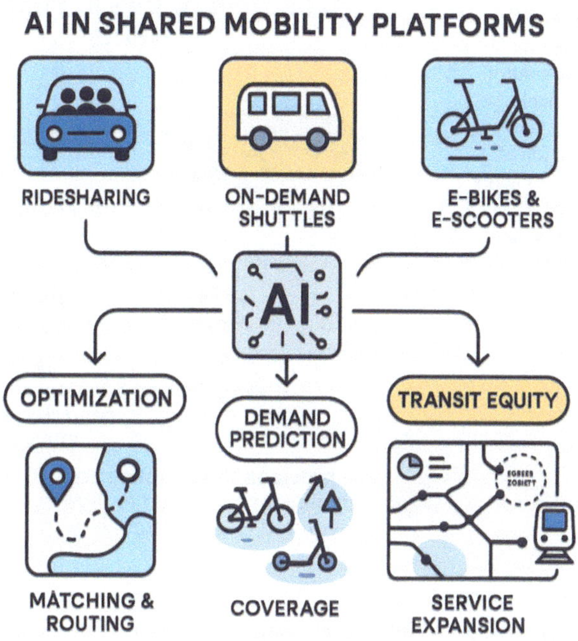

Fig. 7.2 AI powers shared mobility by optimizing matching, routing, and coverage, predicting demand, and expanding services to underserved areas, ensuring efficiency, equity, and accessibility across ridesharing, on-demand shuttles, and micromobility networks

7.3 Smart Charging: AI for Electric Vehicle Infrastructure and Grid Integration

One of the main pillars of transportation decarbonization is the explosive growth of electric vehicles (EVs), but the shift is far more complicated than just switching to electric drivetrains from internal combustion engines. Every EV has the potential to be a distributed energy asset in addition to being a mobile energy consumer (Cheshmehzangi et al., 2025). Energy systems have both new opportunities and difficulties as a result of this dual role, particularly as adoption grows. Uncoordinated charging has the potential to underuse renewable energy, overwhelm local systems, and worsen peak demand if it is not managed carefully. AI provides the adaptive control and predictive intelligence required for EVs to be smoothly integrated into the larger energy environment (See Fig. 7.3).

One of AI's most critical contributions lies in charging demand prediction and infrastructure placement. To predict where and when charging demand will occur, machine learning algorithms can analyze a wide range of variables, including current EV adoption rates, predicted growth, travel patterns, land use data, driver demographics, and even planned infrastructure projects. AI assists planners in avoiding

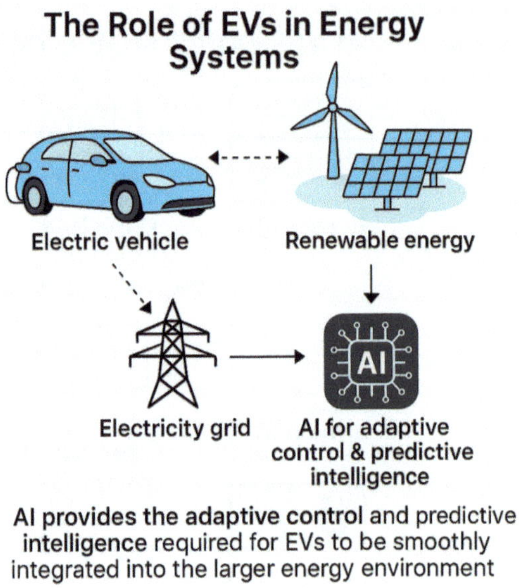

Fig. 7.3 Electric vehicles, renewable energy, the electricity grid, and AI-driven adaptive control work together to enable efficient, predictive, and sustainable integration of EVs into the broader energy system, reducing peak demand and maximizing renewable utilization

both under-provisioning (which causes range anxiety and bottlenecks) and over-investing (which results in underutilized assets) by modelling various scenarios. In the Netherlands, where AI-driven modelling has directed a dense (Mishra et al., 2025), uniformly dispersed charging network that adjusts to changes in EV demand at the neighbourhood level, this predictive approach has been crucial.

For the sake of cost and carbon footprint, charging time optimization is equally crucial. Forecasts of grid load, current energy prices, and the availability of renewable energy sources can all be used by AI systems to plan charging times. AI can maximize the amount of renewable electricity in the battery mix by delaying charging until hours when a wind energy surplus is anticipated, for instance. By taking into account a driver's departure time, average mileage, and battery condition, this optimization may be customized at the user level to guarantee that charging satisfies mobility demands without placing undue load on the grid. Coordinated off-peak charging can dramatically flatten demand curves, negating the need for expensive grid upgrades, as demonstrated by California's smart charging pilots.

The integration of vehicle-to-grid (V2G) systems, which permits bi-directional energy transfer between EVs and the grid, is the next stage in this progression (Mojumder et al., 2022). Here, AI takes on the role of the orchestration engine, deciding when to use renewable energy cycles, when to recharge vehicles, and when to pull electricity from them depending on grid stability requirements and vehicle owners' mobility needs (Arfeen et al., 2020). V2G-enabled fleets have the potential to function as a massive, adaptable energy storage network at scale, reusing extra solar energy throughout the day and reusing it during evening peak hours (Aguilar Domeinguez, 2023). To guarantee that a vehicle's availability for the subsequent journey is never jeopardized by its discharge, AI must make snap decisions. These advantages are further enhanced by fleet-level charge control. Future autonomous fleets, delivery trucks, and buses follow predictable routes and schedules, which gives AI the chance to optimize charging spots and windows. For instance, delivery vans can get top-up charges at distribution hubs in between runs, and electric buses can be charged at depots during midday solar peaks. In order to avoid concurrent demand spikes, AI can also schedule charging sessions for several fleets spread out throughout a metropolis. An early example of how intelligent scheduling might expand beyond individual vehicles is Tesla's integrated strategy, which automatically routes cars to the best charging station and dynamically modifies charging rates.

Real-world implementations demonstrate how EVs, smart grids, and AI have come together to form a new mobility–energy nexus. ElaadNL's AI-managed charging infrastructure in the Netherlands allows for real-time balancing between energy and transportation systems by dynamically adjusting charge rates in response to grid frequency and renewable supply (Van Den Brink & Broos, 2022). Pacific Gas and Electric (PG&E) in California has teamed up with AI companies to predict charging demand throughout its service area, maximizing home charging incentives as well as public charging stations (Lipman & Yuan, 2025). These illustrations show that the true potential of smart charging is found in coordinating thousands or even millions of separate charging events to support grid stability and mobility requirements.

The interaction between automobiles and energy networks is changing as a result of this convergence. EVs are becoming active members of a dispersed, intelligent grid rather than passive consumers of electricity. By bridging the gap between grid dynamics, user behaviour, and renewable variability, AI enables electrification to grow without compromising affordability or dependability. The change is structural rather than merely technological, necessitating new kinds of cooperation between manufacturers, energy suppliers, legislators, and AI researchers.

Charging an EV in this integrated future will be a dynamic, context-aware process that takes into account energy cost, carbon intensity, grid resilience, and individual mobility patterns rather than just plugging in and waiting (Sharif, 2025). This process's intelligence makes sure that every kilowatt-hour is used for the correct reason, at the right time, and in the appropriate location. By doing this, AI transforms the difficulty of integrating EVs with the grid into a tactical advantage, allowing for cleaner transportation and a more adaptable, renewable energy source.

7.4 A New Mobility Intelligence: Governance, Equity, and Decarbonization at Scale

Faster travel, cleaner fleets, and more efficient routes are all potential benefits of AI-powered mobility, but there are also social and political obstacles to overcome. The fundamental problem is data: transportation AI systems rely on massive, ongoing data streams about passenger flows, infrastructure conditions, vehicle movements, and energy networks to function. However, a large portion of this data is kept in silos, either locked within proprietary platforms, dispersed throughout municipal agencies, or protected by commercial operators. Because systems cannot "see" the whole picture required to maximize mobility at scale, the full benefits of AI remain unrealized in the absence of interoperability.

Regulatory slowness makes the issue worse. AI in transportation is developing more quickly than legal frameworks can keep up, leading to ambiguities about privacy safeguards, safety regulations, and the equitable distribution of public space. Budgetary procedures and procurement cycles in the public sector frequently proceed too slowly to keep up with the swiftly evolving technologies. Competitive advantage in the private sector can cause algorithms and decision-making procedures to become opaque, making it challenging for the public or regulators to determine whether systems are serving larger social objectives.

A more subtle but no less important problem is digital exclusion. AI-enabled mobility services, such as dynamic transit routing and app-based ride pooling, frequently presume that customers have smartphones, digital payment methods, and continuous internet connectivity. People who are elderly, live in rural areas, or have poor incomes are left behind since they do not have access to these technologies. AI has the ability to transcend differences, but if it is not purposefully designed for inclusivity, it may unintentionally widen already-existing mobility disparities.

In this new mobility environment, governance needs to strike a balance between responsibility and creativity. The challenge for public transportation organizations and local governments is to include AI without giving up authority over fundamental mobility choices to untransparent, privately held algorithms. Clear transparency criteria are necessary for this, including explainable AI in pricing, scheduling, and routing choices (Garn & Amirghasemi, 2025; Pourroostaei Ardakani & Cheshmehzangi, 2023). Additionally, it calls for public monitoring procedures to make sure mobility AI complies with safety standards, accessibility requirements, and climate goals. Instead of functioning independently, private operators require financial, contractual, or reputational incentives to exchange data, take part in open performance audits, and coordinate planning.

A key component of any AI mobility strategy must be equity (Gao et al., 2025). Peripheral regions are left with aging cars, inadequate transit connections, and increased transportation expenses as a result of electrification and shared mobility initiatives that frequently focus on high-density, wealthy locations where return on investment is fastest. By mapping service gaps, predicting future demand in neglected areas, and optimizing last-mile solutions that connect distant neighbourhoods to major transit corridors, AI can assist in reversing this trend (Sorooshian et al., 2022). However, this will only occur if equality results are incorporated into system design as clear goals rather than being treated as incidental advantages.

Here, open mobility data is a potent facilitator. Building integrated, multimodal systems that operate as a cohesive whole rather than a patchwork of competing services is made possible when anonymized, standardized transport data is shared across public and commercial entities (Wittenberg et al., 2024). This exchange can be formalized through public–private partnerships, which align incentives to serve the public interest and commercial viability. These frameworks have made it possible for AI to coordinate fleets across operators in places like Helsinki and Los Angeles, cutting down on duplication and expanding network reach without compromising competitive innovation.

The more profound reality is that AI is changing the energy logic of cities itself, not just providing a set of tools to modify current transportation systems. The energy and movement fluxes that characterize urban life are rewritten by every AI-coordinated routing choice, charging schedule, and mode transfer. Therefore, decarbonizing transportation involves rethinking the system itself, including its governance, equitable obligations, link to land use, and integration with energy infrastructure, rather than simply replacing technologies in a static system.

In this new environment, AI serves as the glue that binds cars, streets, grids, and people together to form a flexible, responsive whole. Coordination is key to its effectiveness; thousands of small decisions are made every day that, when added together, reduce emissions while enhancing accessibility and quality of life. Policymakers, business executives, and communities face the task of ensuring that this intelligence is used for long-term sustainability, resilience, and equity in addition to efficiency and profit. By doing this, we can design a mobility system that allows everyone to travel in a clean, intelligent manner.

References

Aguilar Dominguez, D. (2023). *The influence of electric vehicle availability on vehicle-to-grid provision within a microgrid* (Doctoral dissertation, University of Sheffield).

Arfeen, Z. A., Khairuddin, A. B., Munir, A., Azam, M. K., Faisal, M., & Arif, M. S. B. (2020). En route of electric vehicles with the vehicle to grid technique in distribution networks: Status and technological review. *Energy Storage, 2*(2), Article e115.

Argyriou, I., & Barry, J. (2021). The political economy of socio-technical transitions: A relational view of the state and bus system decarbonization in the United Kingdom. *Energy Research & Social Science, 79*, Article 102174.

Bhattacharya, A., Songwe, V., Soubeyran, E., & Stern, N. (2024). Raising ambition and accelerating delivery of climate finance. *London: Granthan research institute on climate change and environment, London school of economics and political science.* Available online from: https://cedrec.com/cedrec_images/Raising-ambition-and-accelerating-delivery-of-climate-finance_Third-IHLEG-report.pdf

Van Den Brink, H., & Broos, P. (2022, June). Cyber security challenges in the electric vehicle infrastructure. In *IET conference proceedings CP798* (Vol. 2022, No. 3, pp. 429–432). The Institution of Engineering and Technology.

Cheshmehzangi, A., Allam, Z., He, B. J., & Su, Z. (2025). Climate change efforts vs. the growing popularity of larger private vehicles: A contradiction in motion. *Human Settlements and Sustainability*. https://doi.org/10.1016/j.hssust.2025.07.002

Gao, J., Zuo, F., Yang, D., Tang, Y., Ozbay, K., & Seeley, M. (2025). Toward equitable progress: A review of equity assessment and perspectives in emerging technologies and mobility innovations in transportation. *Journal of Transportation Engineering, Part a: Systems, 151*(1), 03124003.

Garn, W., & Amirghasemi, M. (2025). Transparency of combinatorial optimisations via machine learning and explainable AI. *Annals of Operations Research*, 1–32.

Hellström, T. (2007). Critical infrastructure and systemic vulnerability: Towards a planning framework. *Safety Science, 45*(3), 415–430.

Lah, O. (2017). Decarbonizing the transportation sector: Policy options, synergies, and institutions to deliver on a low-carbon stabilization pathway. *Wiley Interdisciplinary Reviews: Energy and Environment, 6*(6), Article e257.

Lipman, T., & Yuan, J. (2025). Electric vehicle charge management strategies to benefit the California electricity grid. Available online from: https://escholarship.org/uc/item/19n0s7np

Mishra, D. P., Dash, A. K., Behera, S. R., Patro, K. A., & Salkuti, S. R. (2025). AI-driven optimization techniques for electric vehicle charging infrastructure. In *Artificial intelligence for integrated smart energy systems in electric vehicles* (pp. 283–308). Springer Nature Switzerland.

Mittal, S., & Ukkusuri, S. V. (2025). Overcoming data scarcity in transit planning: A novel framework combining machine learning and metaheuristics. *Data Science for Transportation, 7*(1), 4.

Mojumder, M. R. H., Ahmed Antara, F., Hasanuzzaman, M., Alamri, B., & Alsharef, M. (2022). Electric vehicle-to-grid (V2G) technologies: Impact on the power grid and battery. *Sustainability, 14*(21), 13856.

Pourroostaei Ardakani, S. & Cheshmehzangi, A. (2023). *Big data analytics for smart transport and healthcare systems.* Singapore: Springer.

Sharif, M. (2025). *Context-aware optimal resource management in electric vehicle smart2charge* (Doctoral dissertation, Birmingham City University).

Sorooshian, S., Khademi Sharifabad, S., Parsaee, M., & Afshari, A. R. (2022). Toward a modern last-mile delivery: Consequences and obstacles of intelligent technology. *Applied System Innovation, 5*(4), 82.

Wittenberg, D., Paschke, A., Kukuk, A., & Pannek, J. (2024). Information requirements and legal framework for multimodal transport system coordination. *Logistics, 8*(4), 123.

Part III
AI-Enhanced Urban Planning and Climate-Responsive Infrastructure

This section looks at how AI is transforming infrastructure and urban planning to adapt to the needs of a changing climate. AI-powered urban simulations known as "digital twins" give planners the ability to test ideas before putting them into action, optimize green infrastructure, and evaluate environmental effects. By examining materials, heat dynamics, and microclimates, AI also helps climate-responsive design by guiding the creation of more resilient structures and public areas. Predictive maintenance algorithms also keep an eye on vital infrastructure, such as water systems, bridges, and roadways, predicting faults and directing prompt repairs. Smarter, more adaptable cities that can survive climate shocks and maintain long-term urban resilience are fostered by these advancements.

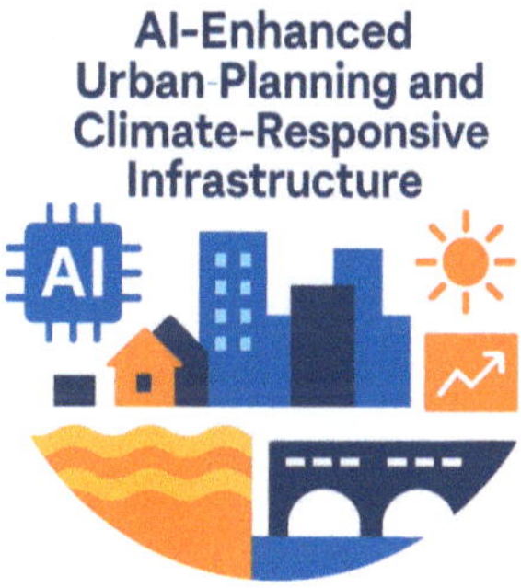

Chapter 8
Living Models: Digital Twins for Climate-Responsive Urban Futures

8.1 Cities in Parallel: The Emergence of Digital Twins in Urbanism

The city, as we know it, is no longer a static entity to be mapped, surveyed, or simply built. Cities are changing into multifaceted, data-driven, self-reflexive dynamic organisms in the era of interconnected systems and escalating climate catastrophes. The digital twin, a living, learning model of the urban environment driven by streams of real-time data and enhanced by AI, is at the front of this progress. Digital twins have transcended their industrial roots and are transforming urbanism itself. Previously, they were only used in advanced manufacturing and aerospace (Li, 2024). What started out as a technique to simulate the performance of a jet engine has evolved into a revolutionary tool for predicting future events, modelling urban complexity, and creating adaptive solutions to problems at the planetary scale.

In its most advanced urban form, a digital twin is more than just a 3D model. It is an advanced cyber-physical system that combines information from social patterns, mobility systems, environmental sensors, municipal infrastructures, and more. Digital twins are animated by real-time feedback loops, in contrast to static tools like Geographic Information Systems (GIS) or even sophisticated Building Information Modelling (BIM) or City Information Modelling (CIM). They react to changing circumstances, learn from the present, and model potential future events. A fundamental change in urban governance from reactive planning to predictive, pre-emptive design is symbolized by this increase in responsiveness.

Even while they are strong in their own right, traditional urban tools mostly rely on photographs. GIS maps transportation flows, demographics, and terrain. BIM facilitates stakeholder and scale coordination in construction. However, neither naturally conveys the energy of a bustling metropolis. On the other hand, live data streams—from satellite imagery, traffic sensors, weather stations, Internet of Things devices,

© The Author(s), under exclusive license to Springer Nature Singapore Pte Ltd. 2025
A. Cheshmehzangi, *AI Agency in Climate Emergency*, Urban Sustainability,
https://doi.org/10.1007/978-981-95-4192-8_8

and even mobile apps—are used to feed digital twins (Ward et al., 2023). AI algorithms then analyze this data to simulate system-to-system interactions, test interventions in silico, and produce insights that change as the city does. Instead of a city model, the end product is a city in parallel—an environment that is constantly present that shadows, mimics, and occasionally even surpasses its real-world equivalent.

Why is this level of intelligence necessary for cities today? due to the fact that their problems are no longer isolated or linear. Climate threats can take many forms: heat waves strain water and power infrastructure; floods cause transportation disruptions and medical emergencies; and smoke from wildfires changes human behaviour and impairs respiratory health. These are not isolated incidents. They are nonlinear, intricately linked, and becoming more unpredictable. The outdated segregated planning and static forecasting models are no longer adequate. Cities need tools that can understand interdependencies in real time, adjust as systems interact, and forecast across aspects that were previously handled independently to manage this complex web of relationships.

This is where AI and digital twins converge as co-architects of the climate-responsive city (Ali et al., 2025; Islam, 2025). Digital twins may now explore generative simulation and continuous learning, going beyond descriptive modelling, thanks to AI. Machine learning algorithms can detect subtle relationships that human planners would miss, such as those between rainfall and groundwater recharge, traffic density and air quality, and weather patterns and energy surges. By predicting how climate scenarios would develop over the urban area, deep learning helps maximize mitigation strategies. Reinforcement learning may evaluate the effects of various design possibilities, test them, and develop the best adaptable solutions (Caputo & Cardin, 2022). Digital twins are essentially turned from passive mirrors into active agents by AI, which enables them to diagnose, simulate, and direct real-time decisions with a level of accuracy and agility never previously achievable.

Nowhere is this fusion of AI and digital twins more urgent than in the climate domain (Cheshmehzangi, 2026). Once considered a sporadic calamity, urban flooding is now a persistent and growing hazard. Cities may simulate runoff paths, detect flood-prone areas beforehand, and implement mitigation methods in real time by using high-resolution digital twins that are fed by rainfall sensors, drainage data, and topographic modelling. Similarly, thermal sensors, land surface data, and demographic overlays can be used to mimic excessive heat, a silent killer made worse by the urban heat island effect. Cities may precisely target cooling measures, such as greening programs, albedo-enhancing materials, or shaded transportation routes, with AI-augmented twins. Through live traffic analysis and behavioural modelling, mobility patterns—which are put under more stress during climate shocks—can be rearranged to enable adaptive rerouting, the strengthening of public transportation, or the integration of micro-mobility in crisis areas.

Intelligent digital twins can be used to monitor and control even infrastructure strain, whether it comes from aging assets, water scarcity, or energy demand. Energy networks, for example, are able to automatically rebalance distribution and predict surges associated with heatwaves. Water systems can identify leaks, predict shortages, and optimize use based on actual consumption trends. To put it briefly, digital

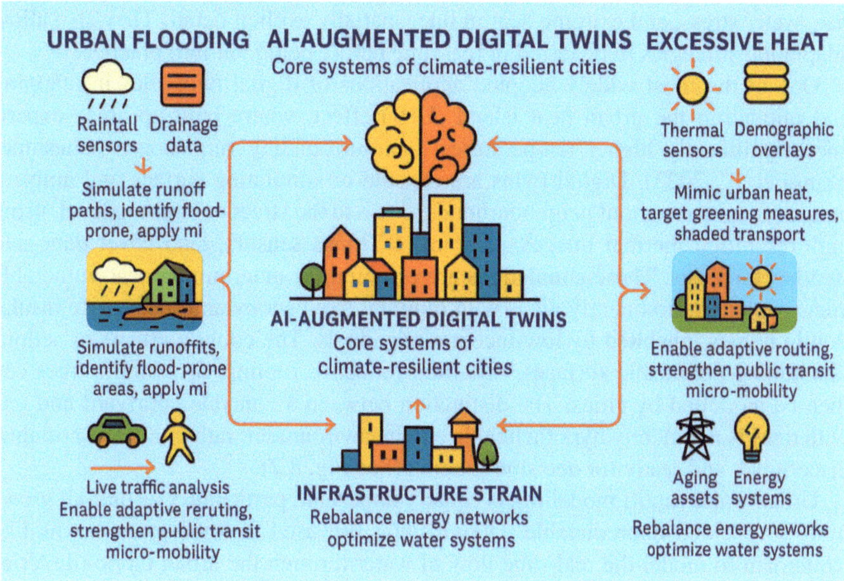

Fig. 8.1 AI-powered digital twins form the core systems of climate-resilient cities, enabling real-time simulation, prediction, and adaptive response to urban flooding, heat stress, mobility disruptions, and infrastructure strain across sectors

twins serve as the core systems of climate-resilient cities, detecting stress, anticipating thresholds for responses, and coordinating cross-sector adaptive measures (See Fig. 8.1).

As this chapter progresses, we will examine how these live models serve as the basis for a new urban epistemology, which views the city as an ecosystem of flows, behaviours, hazards, and opportunities, rather than just being technical tools. Digital twins provide both visibility and foresight in the face of growing climatic unpredictability. They provide us the ability to plan precisely, to experiment without fear of repercussion, and—above all—to take action before the harm is done. Once a work of science fiction, the parallel city is now a moral need.

8.2 Simulating Climate Impacts and Adaptation Strategies

The ability to predict, model, and adjust to climate stressors is becoming essential as cities become more affected by the climate catastrophe. Digital twins have become essential tools for modelling the growing effects of climate change on urban systems, especially when they are driven by AI. These living models, which are far from static visualisations, allow cities to comprehend the effects of air pollution, sea level rise, water stress, and extreme heat in fine, spatially explicit detail. They also allow adaptation strategies to be tested in real time before being put into practice.

One of the most widely adopted applications of digital twins lies in mapping and mitigating the urban heat island (UHI) effect, where built-up areas experience significantly higher temperatures than surrounding rural zones (Cheshmehzangi et al., 2023). Digital twins are capable of simulating surface and ambient temperatures throughout neighbourhoods, down to the street or building level, using high-resolution thermal images, Internet of Things sensors, land cover data, and weather forecasts. These simulations are essential for determining heat-vulnerable areas, which are frequently areas with minimal tree canopy and inadequate insulation in homes inhabited by low-income individuals. The cooling effects of actions like adding permeable surfaces, installing reflective roofing, or planting trees can then be modelled by cities. The distinction between a concrete courtyard and one with trees is not merely hypothetical in a twin environment; rather, it is quantifiable, discernible, and ready for decision-making (See Fig. 8.2).

Urban water runoff modelling is equally important, particularly as rainfall grows more severe and unpredictable. Digital twins are used in flood-prone towns like Rotterdam to model the real-time flow of water through the urban fabric (de Vries et al., 2024). In order to estimate where water will accumulate and when systems will be overloaded, these models include information from drainage infrastructure, soil permeability, elevation, and precipitation forecasts. At the same time, communities can assess how well retention basins, rain gardens, green roofs, and bioswales slow or divert water. This kind of in-silico hydrological testing allows urban planners and emergency services to pre-position resources, reinforce weak infrastructure, and

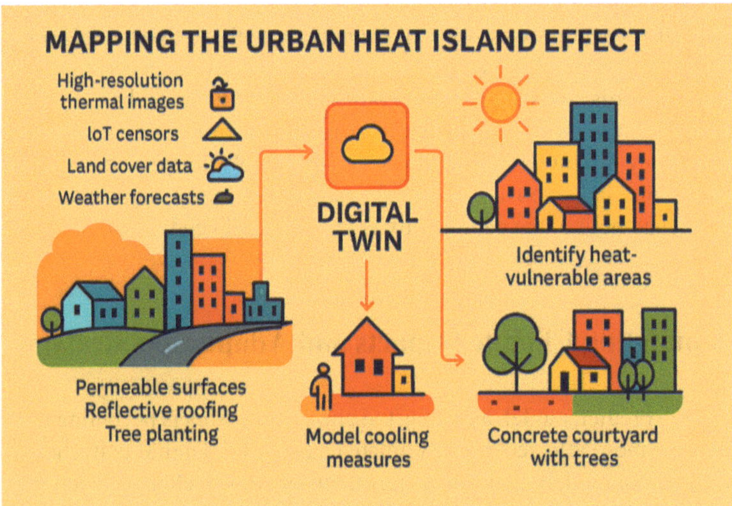

Fig. 8.2 Digital twins help cities combat the Urban Heat Island effect by simulating heat patterns, identifying vulnerable zones, and modelling cooling interventions—enabling data-driven decisions for equitable, sustainable, and climate-resilient urban environments

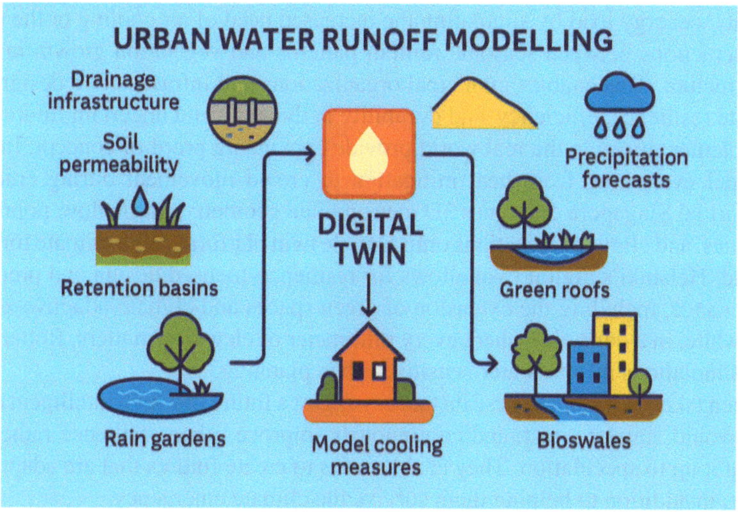

Fig. 8.3 Digital twins enable cities to model urban water runoff in real time, integrating drainage, soil, elevation, and rainfall data to predict flooding, test green infrastructure solutions, and guide adaptive planning for resilient, hydrologically responsive urban development

design future developments that are hydrologically adaptive rather than reactive (See Fig. 8.3).

Due to the combined effects of urbanization and climate change, air quality is also becoming more unpredictable and hazardous. Digital twins can create localized air quality simulations that vary dynamically with time of day, temperature, wind, and movement patterns by ingesting real-time data from traffic sensors, air pollution monitors, and weather forecasts. Cities can use this type of forecasting to predict smog outbreaks, reroute transportation, or alert residents to exposure dangers. These models can also assist with focused public health initiatives in populations that are at risk when they are superimposed with demographic and health data.

Adapting to sea level rise and preventing coastal floods are two of the most important applications of climate simulation. Planners can evaluate long-term sea-level scenarios based on IPCC forecasts and regional subsidence trends using digital twins in low-lying cities like Singapore, which are vulnerable to chronic flooding. Authorities can model how floodgates, seawalls, mangrove replanting, or elevated infrastructure will function under various future scenarios by installing a virtual mirror of the city. To safeguard subsurface transportation and building foundations, Helsinki has also created a climate-adaptive twin that simulates snowfall patterns and rising groundwater levels. These twins calculate the advantages and disadvantages of proactive protection in addition to visualizing damage.

These simulations' AI-enhanced capacity to manage enormous datasets, find hidden correlations, and perform iterative scenario planning is what makes them particularly potent. AI can examine the degradation of materials used for rooftops or roadways under frequent heat waves. It can mimic the effects of a protracted drought on a city's energy load by simulating the increased need of air cooling or the potential interactions between seasonal rainfall patterns and vegetation growth in green infrastructure. This enables municipal organizations and infrastructure designers to evaluate resilience, efficiency, and durability in dynamic and uncertain futures.

Implementations in the real world provide convincing proof of concept. In order to model everything from heat mitigation to crowd movement during emergencies, Virtual Singapore combines 3D metropolitan geometry, traffic flow, population dynamics, and climate projections onto a single twin platform. With climate foresight in mind, Helsinki's digital twin allows for resilience-focused design and predictive maintenance, including the extension of green spaces and climate-adaptive zoning. Meanwhile, in a delta city where every millimeter of elevation matters, Rotterdam's flood simulations inform water-sensitive urban planning.

When taken as a whole, these instances suggest a future in which intelligent digital twins would simulate, test, and continuously improve urban resilience rather than leaving it up to speculation. They enable cities to create futures that are adaptive by default, in addition to helping them survive the climate emergency.

8.3 Integrating Nature and Infrastructure in Virtual Space

Integrating nature into urban infrastructure is becoming a survival necessity as cities enter an era of deep adaptation, not a luxury or aesthetic preference (Scott et al., 2016). Green and blue infrastructure systems—living networks of vegetation, water bodies, soils, and ecosystems woven into the urban fabric—are becoming increasingly important in the design of resilient cities as a result of biodiversity loss, rising temperatures, deteriorating air and water quality, and social fragmentation. With the help of real-time data and AI-enhanced analytics, digital twins provide a previously unheard-of chance to model, simulate, and optimize these hybrid infrastructures, guaranteeing that technology and nature work together rather than against each other (See Fig. 8.4).

The ability of digital twins to simulate the combined advantages of green and blue infrastructure in terms of the environment, society, and economy lies at the core of this change. For instance, trees and urban forests do more than just look good; they also control microclimates, absorb pollutants, lessen stormwater runoff, and enhance mental health. Wetlands lower the risk of flooding, store carbon, and filter water. Permeable streets promote groundwater recharging and stop heat accumulation. These ecosystem services can be measured, plotted, and evaluated against various urban and climatic conditions using digital twins. Planners can use simulations to quantify the effects of green roofs on indoor cooling needs or the ways in which park accessibility promotes social cohesiveness at the neighbourhood level.

This capacity for performance-based simulation is what sets digital twins apart from conventional design tools. It is possible to model regenerative infrastructure, such community gardens, urban wetlands, and bioswales, not just for their spatial or visual integration but also for their role in urban metabolism. How much water, for

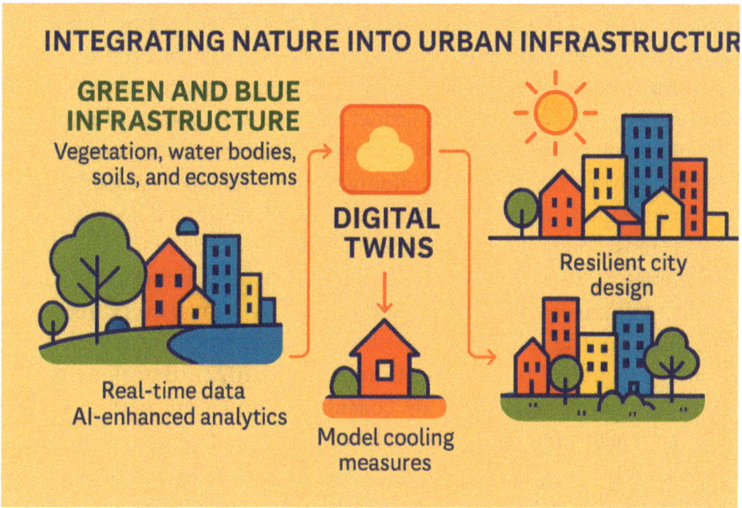

Fig. 8.4 Digital twins empower cities to integrate green and blue infrastructure by simulating nature-technology interactions. Using real-time data and AI, planners can optimize ecosystems—vegetation, water, and soils—for climate resilience, biodiversity, and healthier urban living in an era of deep adaptation

example, can a green roof hold during a storm that occurs once every 100 years? In a crowded street canyon, how much of a temperature drop may be anticipated from a tree canopy? What is the possibility for street-side plants to sequester carbon during a ten-year period? These aren't hypothetical inquiries. The answers are available before any concrete is poured or trees are planted thanks to virtual models and sensor data that are fed into AI systems. These models can also be used iteratively to test cost–benefit ratios, compare designs, and improve treatments to maximize ecological performance.

The capacity of digital twin environments to facilitate cross-sectoral coordination—modelling several infrastructures concurrently to take interdependencies into account—is one of its main advantages. In addition to density and design, housing developments can be evaluated for their exposure to urban heat, shading from green corridors, integration with mobility networks, and availability of blue spaces. Transport planners can find areas where shaded bus stops or green tram corridors could provide relief by superimposing heat vulnerability maps with traffic movements. Health agencies can identify hotspots for respiratory illnesses associated with poor air quality or a lack of green space by using environmental data from twins. Digital twins offer a shared spatial intelligence platform where decisions made in one area are mirrored and optimized in other areas, eliminating the need for siloed planning.

This integration is not only possible but also clever due to AI's optimization capabilities. In order to determine the best design for a certain goal, such as social equity, biodiversity, or cooling, algorithms can model an infinite number of different combinations of space layouts, material kinds, and planting patterns. A generative AI model, for instance, can suggest street designs that preserve sightlines and pedestrian movement while optimizing tree coverage. It can locate unused rooftops that would be suitable for solar-green roof hybrids, which would balance stormwater management and energy generation. With consideration for soil type, rainfall patterns, and long-term maintenance expenses, it can even suggest context-sensitive materials for permeable pavements. Adaptive intelligence of this type replaces guesswork in infrastructure design with performance-driven, evidence-based decision-making.

One of the most compelling real-world illustrations of this approach is Barcelona's green infrastructure strategy, which is being increasingly informed by digital twin technologies. Initiatives have been started by the city to map every tree, green area, and water feature. The information is then fed into virtual platforms that mimic the "green connectivity" of the city (Combs et al., 2022). Barcelona may prepare for climate adaptation holistically by considering urban greenery as a networked system rather than discrete areas. For instance, simulations demonstrate how a green corridor extension might lessen heat stress along public transportation lines or how a continuous loop connecting local parks enhances walkability and supports pollinator species. In order to examine the ways in which green infrastructure interacts with current utilities and public services, the digital twin also incorporates data from the energy, transportation, and health sectors.

Digital twins essentially offer a spatial laboratory where infrastructure and environment are not only co-located but also co-evolving. They assist cities in moving past the traditional division between the natural and built environments. By doing this, they unveil a future where infrastructure flows, breathes, shades, and heals—a regenerative urbanization that aims to promote rejuvenation as well as resilience.

8.4 Toward Participatory, Predictive, and just Urban Twins

A major adjustment is needed from considering digital twins as high-tech planning tools to accepting them as civic infrastructures, as they become increasingly important in cities' understanding and response to climate change. Urban digital twins have the potential to revolutionize the way we plan, manage, and live in urban futures (Pourroostaei Ardakani & Cheshmehzangi, 2024). This potential extends beyond their technical prowess or real-time responsiveness. However, we must address the institutional, ethical, and social issues that come with their quick adoption if we are to fulfill this potential.

Data governance, privacy, interoperability, and accessibility are some of the most urgent problems. Digital twins are dependent on large and frequently sensitive data streams that are gathered from ambient sensors, private mobility apps, public infrastructure, and occasionally even individuals. Who is the owner of this data? Who manages its flow? Furthermore, how can cities make sure that this data is not utilized for surveillance or commercial exploitation but rather for the sake of all? Digital twins run the potential of becoming opaque, centralized systems that exacerbate injustice and mistrust rather than promoting openness and cooperation in the absence of strong governance mechanisms.

This is the point at which the twin needs to be built with inclusion and participation from the beginning. The most potent urban twins are those that let communities to access layers of scenario-building, visualization, and feedback rather than those that are restricted to proprietary platforms. Digital twins can develop into instruments of co-design and democratic planning by enabling locals to observe the effects of climate actions on their neighbourhoods, offer suggestions, and even model their own alternatives. Open twin platforms, for instance, might let residents to investigate how putting trees on their street might lessen heat stress in the area or how zoning changes might impact access to green space, walkability, and air quality. By bridging the gap between the tangible and the technical, these participatory layers transform digital abstraction into civic empowerment.

However, involvement is insufficient on its own. Hard ethical considerations about who the twin is for—and who might be excluded—must also be asked. The assumptions that AI-driven simulations encode and the data they receive influence them. The digital twin may covertly reproduce current disparities if these inputs favour wealthy areas with superior sensors or predict results using limited economic models. Whose voice is absent? Whose dangers are ignored? The algorithm is being programmed with whose priorities? Digital twins need to be examined, challenged, and co-owned by the communities they represent as well as engineers and planners to guarantee justice in climate adaptation.

Digital twins have potential for the future as they evolve from technological assets to urban infrastructure. They must be funded as shared goods, governed in the public interest, and built with transparency, accountability, and equity at their center, much like public parks, libraries, or transit systems. This calls for a rethinking of the production and application of urban knowledge as a collaborative process toward resilience rather than a top-down exercise in forecasting. Cities can create digital twin ecosystems that act as venues for inclusive innovation, experimentation, and foresight with the correct collaborations, legislation, and public discussions.

In conclusion, digital twins become more than just digital representations of the city when they are directed by ethical intelligence and community involvement; they become catalysts for urban foresight (Fontes & Dubey, 2024). They enable us to see alternatives, predict danger, and model complexity. More significantly, they ask us to do so in a fair, open, and cooperative manner. Their most daring pledge is to contribute to the creation of a future that is not only data-driven and adaptive, but also liveable and equitable for everyone in an era characterized by social fragmentation and climatic unpredictability.

References

Ali, E., Mansour, A., Mohammed Abdelkader, E., Elshaboury, N., & Zayed, T. (2025). Digital twin for climate resilience: Transforming smart cities for a sustainable future. *The International*

Archives of the Photogrammetry, Remote Sensing and Spatial Information Sciences, 48, 139–145.

Caputo, C., & Cardin, M. A. (2022). Analyzing real options and flexibility in engineering systems design using decision rules and deep reinforcement learning. *Journal of Mechanical Design, 144*(2), Article 021705.

Cheshmehzangi, A. (2026). *Generative AI-powered urban digital twins: Pioneering environmental solutions for sustainable intelligent cities*. Springer.

Cheshmehzangi, A., He, B. J., Sharifi, A., & Matzarakis, A. (2023). Climate change, cities, and the importance of cooling strategies, practices, and policies. In *Climate change and cooling cities* (pp. 2–19). Springer Nature Singapore.

Combs, M. A., Kache, P. A., VanAcker, M. C., Gregory, N., Plimpton, L. D., Tufts, D. M., Fernandez, M. P., & Diuk-Wasser, M. A. (2022). Socio-ecological drivers of multiple zoonotic hazards in highly urbanized cities. *Global Change Biology, 28*(5), 1705–1724.

Fontes, C., & Dubey, R. K. (2024). Urban futures: possibilities and challenges for ethical virtual cities. *The Elgar Companion to Applied AI Ethics*, 290–312.

Islam, F. S. (2025). Artificial Intelligence-powered Carbon Market Intelligence and Blockchain-enabled Governance for Climate-responsive Urban Infrastructure in the Global South. *Journal of Engineering Research and Reports, 27*(7), 440–472.

Li, X. S. (2024). *Building digital twin metaverse cities: Revolutionizing cities with emerging technologies*. Springer Nature.

Pourroostaei Ardakani, S. & Cheshmehzangi, A. (Eds.). (2024). *Digital twin computing for urban Intelligence*. Springer.

Scott, M., Lennon, M., Haase, D., Kazmierczak, A., Clabby, G., & Beatley, T. (2016). Nature-based solutions for the contemporary city/Re-naturing the city/Reflections on urban landscapes, ecosystems services and nature-based solutions in cities/Multifunctional green infrastructure and climate change adaptation: Brownfield greening as an adaptation strategy for vulnerable communities?/Delivering green infrastructure through planning: Insights from practice in Fingal, Ireland/Planning for biophilic cities: From theory to practice. *Planning Theory & Practice, 17*(2), 267–300.

de Vries, M., Vinke-de Kruijf, J., & Augustijn, D. D. (2024). Design of a value-based and forward-looking framework to assess flood resilience. In *International conference on resilient systems, ICRS 2024* (p. 55). ETH Zurich.

Ward, R., Choudary, R., Jans Singh, M., Roumpani, F., Lazauskas, T., Yong, M., Barlow, N., & Hauru, M. (2023). The challenges of using live-streamed data in a predictive digital twin. *Journal of Building Performance Simulation, 16*(5), 609–630.

Chapter 9
Designing with Climate: AI for Climate-Responsive Architecture, Urban Form, and Microclimate in a Warming World

9.1 Architecture in the Age of Climate Adaptation

Climate and architecture have long been in conversation. Vernacular architecture evolved as a personal reaction to the environment, from the thick mud walls of Mesopotamian dwellings to the sloping roofs of Southeast Asian buildings that were tailored to the monsoon. Local materials, spatial orientation, thermal tactics, and passive ventilation techniques were all infused with architectural expertise in each locale (Roaf & Cheshmehzangi, 2025; Toroxel & Silva, 2024). These structures served as active intermediaries between humans and their natural environments in addition to being manifestations of culture and identity. Climate-adapted architecture has historically thrived via long-term observation and lived experience rather than abstract theory—a gradual, generational process of adjusting the built environment to the cycles of nature.

However, this gradual build-up of design intelligence is being challenged in the Anthropocene. The climate is become a destabilizing factor rather than a dependable background. Cities and structures are currently facing unprecedented heat waves, more intense storms, unexpected floods, increasing sea levels, and material deterioration due to UV radiation and heat stress (Consalo, 2022). What used to be uncommon is now commonplace. What used to be seasonal is now unpredictable. Static weather, plentiful energy, and dependable materials—the same presumptions that supported contemporary architecture—are eroding. The built environment must learn to not only withstand change but also to anticipate and adapt to it in the age of climate adaption.

Traditional design tools are finding it difficult to keep up in this new era. The intricacy of future climatic scenarios is frequently not captured by static models and simulation software (Cheshmehzangi et al., 2022). Evaluations of building performance frequently use historical averages that are out of date. Furthermore, the iterative process of architectural prototyping is still expensive, time-consuming, and unresponsive to the rapid changes in the environment. Without a clear idea of the

© The Author(s), under exclusive license to Springer Nature Singapore Pte Ltd. 2025

A. Cheshmehzangi, *AI Agency in Climate Emergency*, Urban Sustainability,
https://doi.org/10.1007/978-981-95-4192-8_9

climate in ten, twenty, or fifty years, architects are compelled to make important design decisions about materiality, form, orientation, and energy systems. Despite being the cornerstone of innovation, trial-and-error methods are no longer adequate in a world where climate dangers are increasing.

At this point, AI transforms from a tool into a design collaborator. AI presents a completely new way of thinking that is focused on the future and rich in data. AI can assist architects in comprehending the micro and macro climate realities that a building or district will encounter by combining large environmental datasets, such as thermal maps, hydrological models, urban heat island profiles, sun exposure, and wind behaviour. Regional climate projections can be analyzed using machine learning algorithms, which can then be used to generate useful design suggestions. In just a few seconds, generative algorithms may simulate thousands of different spatial configurations to maximize material efficiency, airflow, shade, and light.

Form can now follow foresight, whereas in the past it followed function. Rapid prototyping of resilience as well as structure is made possible by AI. Buildings can be tested against a variety of future situations using evolutionary design methods, such as heatwaves in 2035, rainfall patterns in 2040, grid stress scenarios, or material deterioration timeframes (Nabaneeta, 2022). Buildings can now be constructed to adapt to a variety of anticipated futures rather than just one environment. This is simulation as imagination rather than just prediction, enabling machine accuracy to enhance human creativity.

Once based on context sensitivity and passive design, climate-responsive architecture is currently seeing a renaissance thanks to computational intelligence. The fundamental ideas are still the same: employing thermal mass, increasing natural ventilation, minimizing heat gain, aligning structures with the direction of the predominant winds, and utilizing the landscape as a thermal buffer. However, these concepts are now improved, customized, and site-specific in ways that were unthinkable in the past. Building envelopes can be customized by AI using exact topography and meteorological data (Mirzabeigi, 2024). It can suggest recycled or bio-based materials for thermal performance under anticipated salinity or humidity conditions in addition to sustainability. It can offer advice on creating skins that respond dynamically to external conditions or shading devices that change with the angles of the sun (Fig. 9.1).

Crucially, AI enhances an architect's vision rather than taking its place. It makes it possible for climate data and artistic intent to be choreographed more closely. The end effect is building that is both futuristic and archaic, driven by machine learning's analytical capabilities but being rooted in the adaptive logics of vernacular design. From opposing environmental conditions to working with them, it encourages a change from planning for climate to creating with climate.

The role of architecture grows in response to the climate catastrophe. Buildings are no more inert objects; in uncertain futures, they are habitats of human dignity, agents of adaptation, and buffers against climate extremes. And in this enlarged capacity, AI serves as an agent rather than just assistance, directing shape, behaviour, and substance in the direction of resilience (Rane et al., 2024). Future architecture needs to be regenerative, responsive, and anticipatory. This vision becomes urgent and actionable when machine intelligence is included into the design process.

9.2 Microclimates, Morphology, and AI-Driven Environmental Analysis

The success of a structure or urban area in the age of climatic instability depends increasingly on how well it interacts with the surrounding microclimate. The finer details of solar exposure, wind flow, humidity, and shade affect whether a public square is useable in the summer or whether a structure maintains its thermal comfort without the need for mechanical cooling, even though macro-level climate strategies are still crucial. AI is revolutionizing the architectural and urban design process at this scale, where design and lived experience collide. Designers can now decipher the subtle dynamics influencing microclimates and respond with forms, materials, and layouts that are exactly adapted to place thanks to AI-driven environmental analysis.

Massive amounts of site-specific environmental data are first processed by AI. A site's thermal and spatial behaviour is influenced by a number of parameters, including the amount of solar radiation received throughout the year, hourly wind vectors, shading from surrounding plants or structures, thermal albedo of surface materials, slope orientation, and water flow patterns (Berry et al., 2013). Some of this analysis was made possible by traditional methods, but they were sluggish, labour-intensive, and had limited ability to synthesize complicated, interconnected factors. These linkages may now be mapped in real time and at high resolution thanks to AI models that have been trained on local datasets and environmental physics.

The integration of remote sensing data, GIS (Geographic Information Systems), and LIDAR (Light Detection and Ranging) is essential to this evolution (Favorskaya & Jain, 2017). AI systems can create precise digital terrain models that capture the constructed features and fine curves of a place thanks to LIDAR's millimeter-precision three-dimensional spatial information. Land use, vegetation, hydrological, and demographic data are layered onto these models using GIS. Long-term climate trends, reflectivity profiles, and thermal imaging are all provided by remote sensing using satellites or drones. Rich environmental maps that show the microclimate zones within a neighbourhood or city block are produced when these datasets are put into AI systems, especially convolutional neural networks and spatial regression models.

These observations go well beyond scholarly curiosity. They are radically changing how design approaches outdoor comfort and environmental performance. AI systems are capable of simulating thousands of neighbourhood design iterations, examining the effects of changing a building's height or orientation on nearby public spaces' shading and wind flow. Based on inputs such as ground materials, tree cover, and nearby heat sources, they are able to estimate pedestrian-level outdoor thermal comfort. This enables planners to pinpoint areas of discomfort before building ever starts and address them with passive cooling techniques, porosity, and optimal spatial arrangements.

Local responsiveness has long been a key component of passive design. This reactivity is proactive, systemic, and scalable thanks to AI. Building orientation can now be adjusted to account for shifting wind patterns anticipated over the coming decades, in addition to the dominant winds. AI algorithms that prioritize energy balance and user comfort can lead parametric designs for building envelopes that are tuned to absorb or deflect radiation at different times of day or season. Previously challenging to model, complex forms can now be evaluated for reflected solar gain, envelope heat retention, and convective airflow incredibly quickly and accurately. For architects and planners, AI has essentially opened up a new level of climate literacy by transforming modelling into an artistic endeavour and data into intuition.

Applications in the real world are already showing how beneficial this strategy is. AI-driven microclimate modelling is being used to create shaded urban corridors that absorb cooling breezes (Chen et al, 2025) while diverting peak solar loads in Dubai, where excessive heat and solar exposure pose a barrier to liveability. AI-powered parametric software allows for quick testing of overhang designs, materials, and façade perforations, assisting in the reduction of urban heat without compromising aesthetics. AI is utilized to model the cooling effects of vegetation and to optimize building orientation around the dominant monsoon winds in Singapore, a city known for its seamless integration of landscape and climate. This supports adaptive zoning and real-time environmental scoring in the city's Green Mark planning system. In southern Europe, where ancient urban fabrics must adapt to hotter summers, AI technologies are assisting the retrofitting of dense neighbourhoods through smart shade, ventilation corridors, and surface material changes—all based on specific environmental modelling.

A new design ethic—one that views climate as a collaborator rather than a limitation—emerges because of these applications. AI-enhanced microclimate-responsive design holds promise for cities that are more expressive of their global position and more liveable in harsh environments. Creating habitats that breathe, buffer, and react intelligently is more important than merely controlling the temperature or using less energy. This combination of creative architecture and environmental analysis could be the foundation of future-proof design as the climate crisis worsens.

9.3 Material Intelligence and Heat-Resilient Urban Fabric

The materials we use in construction are no longer neutral elements as the climate crisis worsens; rather, they actively influence thermal performance, embodied carbon, and long-term resilience. How cities absorb, store, reflect, or discharge heat depends on the materials used, ranging from the scale of individual structures to large urban districts. Designing for climate resilience necessitates a new sort of intelligence in light of rising temperatures, protracted heatwaves, and extreme humidity. This intelligence must be able to assess materials' environmental impact as well as their behaviour under stress and contribution to liveability. AI is becoming a key instrument in this shift, allowing planners, engineers, and architects to create low-carbon, long-lasting urban fabrics that are also adaptable, regenerative, and thermally sensitive.

AI is revolutionizing the selection, comparison, and use of materials in climate-sensitive design. Historically, availability, cost, and aesthetic preference influenced material specification, with sustainability or thermal requirements added later. A wide range of criteria, such as embodied carbon, thermal conductivity, moisture resistance, lifecycle cost, albedo, and degradation rates under various climate futures, can now be evaluated by AI platforms on a vast database of materials, including novel composites, recycled content, and bio-based alternatives. Which materials can withstand high heat cycles, withstand corrosion from salt, or retain reflectivity over time can be determined by machine learning algorithms trained on real-world performance data. This essentially ingrains foresight into the very fabric of the built environment by enabling architects to use materials that are both practical in the present and climate resilient in the future (Fig. 9.2).

Simulating the behaviour of materials under stress from heat and humidity is one of the most potent uses of AI. For instance, AI-enabled parametric modelling techniques can be used to assess the expansion, contraction, degradation, and emissivity changes of façade materials after years of exposure to adverse weather. Also, AI can forecast the behaviour of porous materials in situations with varying humidity, the failure of sealants due to UV deterioration, and the interactions of novel bio-composites with high moisture content. Smarter detailing, better construction sequencing, and more precise performance guarantees are made possible by this simulation capabilities. These features are essential for buildings that need to survive harsher environmental realities without becoming more mechanically dependent.

AI is essential to urban heat island (UHI) abatement measures in the larger metropolitan context. AI algorithms may locate hotspots in a city and suggest material-based remedies by examining high-resolution thermal images, surface temperature data, land use patterns, and radiation absorption rates. Heat-dissipating surface treatments, vegetated roofs and facades, permeable pavements, and reflective coatings can all be placed exactly where they are most needed. AI models can mimic how permeable materials lower ground temperature and stormwater pooling, how changes in surface albedo impact pedestrian comfort, or how urban vegetation interacts with building design to generate microclimate oasis. Cities can shift from general policies to highly focused and successful thermal plans because to this granular, predictive intelligence (See Fig. 9.3).

One of the most fascinating areas where material science and AI meet is the creation of thermally adaptive façades. Intelligent façades use phase-change materials, dynamic shading, or shape-shifting components to adapt to temperature, solar gain, or occupancy patterns. By calibrating actuation triggers, modelling thousands of design variations, and guaranteeing that the façade maintains comfort and energy efficiency under a variety of circumstances, AI helps optimize these systems. AI may, for instance, predict the cooling potential of vented façade systems in Tokyo's humid summers or identify the most effective design for a responsive brise-soleil in Brisbane's subtropical environment. These smart skins are climatic interfaces that act

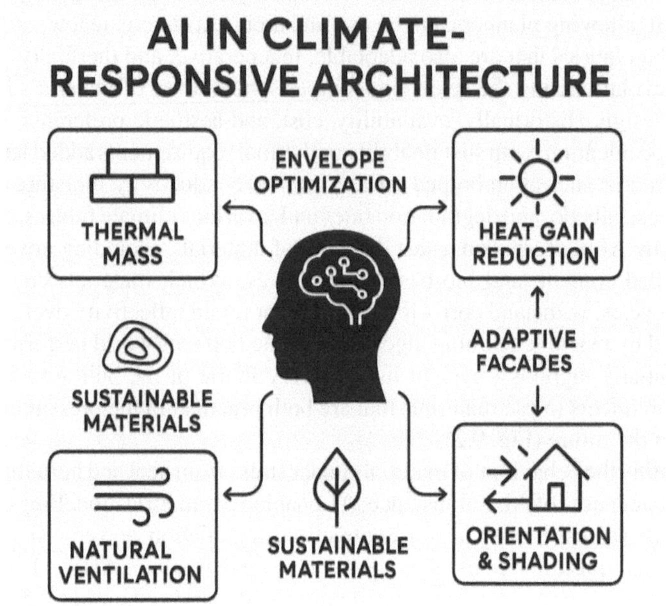

Fig. 9.1 AI enhances climate-responsive architecture by optimizing traditional strategies such as thermal mass, natural ventilation, and orientation. It integrates sustainable materials, adaptive facades, and envelope optimization to create designs tailored to site-specific conditions, improving energy efficiency, comfort, and resilience in the face of changing climate patterns

as a mediator between inhospitable interior conditions and harsh exterior conditions, making them more than just technological innovations.

Examples from Japan and Australia show how this intelligence is being used in real-world situations. Architects are employing AI to model the relationship between materials and site-specific climatic data in Australia's subtropics, where extreme heat, humidity, and sun exposure pose special design issues (Wu et al., 2025). High-emissivity reflective metal cladding, shaded rooflines, and thermally responsive glazing are all features of projects in Darwin and Brisbane that have been optimized using AI techniques that forecast performance across several decades of heat stress. Bio-adaptive building skins that take inspiration from natural heat-regulation techniques are becoming popular in Japan. AI algorithms have been used to simulate the movements of pinecones, insect wings, and plant leaves, creating façades that change their shape in response to changes in temperature. Both biomimicry and data-driven calibration influence these systems, with AI directing each panel's response to surface temperature, wind speed, and solar angle.

In each of these instances, AI enables rather than just helps. It provides urbanists and designers with a fresh perspective on the interplay of form, material, and

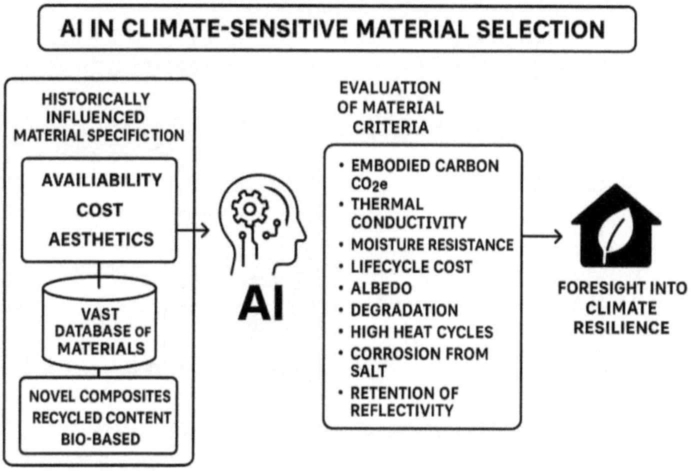

Fig. 9.2 This diagram shows how AI transforms climate-sensitive material selection, evaluating embodied carbon (CO_2e), durability, cost, and performance to identify sustainable, resilient options—integrating foresight into architecture for present functionality and future climate adaptability

climate. It enables the built environment to respond intelligently and purposefully by giving meaning to the invisible forces of heat, moisture, and radiation. Material intelligence won't be an option as cities get hotter, denser, and more susceptible to severe weather. AI makes sure that this intelligence is ingrained in design thinking as well as products, creating an urban fabric that protects its users, cools itself, and revitalizes its surroundings. This is the beginning of materials with a future, not the future of materials.

9.4 Ethics, Aesthetics, and the Future of Climate-Responsive Design

The duty of architecture and urban design increases as the climate crisis changes the air above our houses and the earth underneath our communities. Climate-responsive design is the cornerstone of moral, just, and clever spatial production in a warming world; it can no longer be viewed as a specialized field or a technical add-on. Our current capabilities, especially AI, enable a level of environmental awareness and reactivity never before possible. However, such authority entails the need to inquire: where, whose comfort is valued, and who gains from this intelligence?

In climate design, equity is about fairness, not merely thermal comfort or energy efficiency. The areas in many cities with the least access to adaptable design are also the ones most subject to environmental risks, inadequate ventilation, and heat.

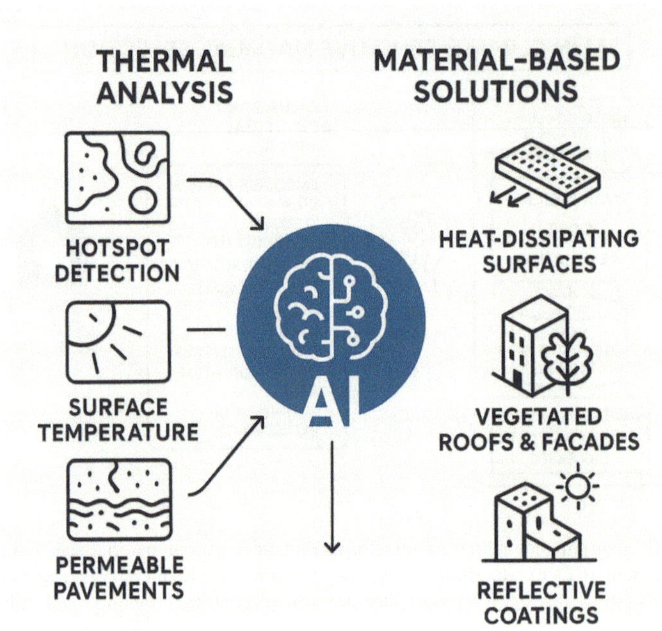

Fig. 9.3 This diagram shows how AI tackles urban heat island effects through precise thermal analysis and targeted material-based solutions. By detecting hotspots, mapping surface temperatures, and evaluating permeable pavements, AI guides the strategic placement of heat-dissipating surfaces, vegetated roofs, and reflective coatings, creating cooler, more resilient metropolitan environments

Low-income neighbourhoods, informal settlements, and outdated public housing sometimes lack the ventilation, shade, and insulation that more affluent regions have. AI-enhanced design runs the risk of exacerbating the very disparities it should help address if it only benefits those who can buy the newest technologies (Dinker, 2024). Policy, funding, and access mechanisms that guarantee these breakthroughs benefit the people, not just a select few, must be incorporated into the future of climate-responsive design.

The harmony between artistic expression and environmental performance is equally crucial. Architecture has always been more than just a place to live; it is a representation of identity, narrative, and memory in space. The difficulty lies in enhancing creative intuition rather than substituting it with algorithms. Instead of reducing beauty to a performance indicator, climate-responsive design should foster new kinds of beauty that are based on environmental discourse. This integration can be accelerated by AI, which enables architects to create designs that are both locally rooted and globally conscious, test expressive forms against energy goals, and combine computational logic with vernacular techniques. It invites the coexistence of meaning and performance in a new poetics of place.

This raises profound questions about authorship in the age of machine intelligence. What happens to the designer's voice as AI plays a bigger role in creating materials, systems, and forms? Rather than opposing technology, the solution is to reinterpret the architect's function as a curator of both artificial and human intellect. Instead of controlling design, AI should work with it to provide options, simulations, and insights while allowing room for morality, creativity, and values. This mix will produce the best climate-responsive designs—data-informed, yes, but also culturally and sympathetically sensitive.

Participatory design procedures and open-access environmental data are essential to this future. Communities are empowered to take charge of their surroundings when they can perceive and comprehend the microclimates in which they live and when they are able to use accessible design tools to co-shape their streets, buildings, and public areas. AI platforms can facilitate communal decision-making, local adaption, and citizen design provided they are made clear and easy to use. In addition to being clever, climate-responsive design needs to be inclusive and based as much on lived experience as it does on forecasting.

Climate-responsive design is now a moral and technological necessity in this period of global instability, not just an aesthetic preference. In a setting that is becoming more hostile, it is how we protect life, culture, and dignity (Sarker and Hossain, 2025). And rather than replacing the human spirit of design, AI is enabling us to tackle this task with unprecedented clarity, accuracy, and inventiveness. It is an ally of the designer, not a substitute, in imagining more equitable, cool, and responsive futures to the shifting cycles of the planet. The way forward is hybrid, adaptable, and incredibly human—it is neither entirely analogue nor entirely digital. And the architecture of the future starts in that hybrid area.

References

Berry, R., Livesley, S. J., & Aye, L. (2013). Tree canopy shade impacts on solar irradiance received by building walls and their surface temperature. *Building and Environment, 69*, 91–100.

Chen, T. Y., Huang, C. S., & Sung, W. P. (2025). Improving summer outdoor comfort in metropolitan park: A data-driven approach using AI, experimental and design analysis. *Journal of Measurements in Engineering*. Available from: https://www.extrica.com/article/24615

Cheshmehzangi, A., Butters, C., Cheshmehzangi, A., Dawodu, A., Jin, R., & Mangi, E. (2022). Urban design in the age of climate change: Paradigms and directions. *Design for vulnerable communities* (pp. 133–154). Springer International Publishing.

Consalo, K. (2022). Vulnerable populations: Climate change and extreme weather threats facing urban communities. *Chi.-Kent J. Env't Energy L., 11*, 1.

Dinker, N. (2024). Artificial intelligence and inequality: Examining the social divides created by technological advancements. *International Journal of Innovations in Science, Engineering and Management*, 228–236.

Favorskaya, M. N., & Jain, L. C. (2017). *Handbook on advances in remote sensing and geographic information systems*. Springer.

Mirzabeigi, S. (2024). *Integrated building envelope assessment towards automation of energy retrofits: Drone-based data acquisition, automated thermal anomaly detection, and workflow development* (Doctoral dissertation, State University of New York College of Environmental Science and Forestry).

Nabaneeta, S. (2022). *Impact assessment and adaptability of a building within a SPEN in a changing Climate: Robustness of BEP systems in terms of future climate, a study on impact of climate change using future climate model* (Master's thesis, NTNU).

Rane, N., Choudhary, S., & Rane, J. (2024). Artificial intelligence for enhancing resilience. *Journal of Applied Artificial Intelligence, 5*(2), 1–33.

Roaf, S., & Cheshmehzangi, A. (2025). Persian vernacular buildings—lessons on resilient design. In *Persian vernacular architecture: Lessons from master builders of Iran on climate resilient design* (pp. 1–12). Springer Nature Singapore.

Sarker, M. M. K., & Hossain, M. S. (2025). Value of climate responsive design in respect of changing environment of nearest future. *International Journal*, *2*(2).

Toroxel, J. L., & Silva, S. M. (2024). A review of passive solar heating and cooling technologies based on bioclimatic and vernacular architecture. *Energies, 17*(5), 1006.

Wu, X., Wang, Y., Deng, S., & Su, P. (2025). Climate-responsive design of photovoltaic façades in hot climates: Materials, technologies, and implementation strategies. *Buildings, 15*(10), 1648.

Chapter 10
The Vigilant City: AI for Predictive Maintenance and Climate-Resilient Infrastructure

10.1 Urban Infrastructure Under Pressure: Age, Risk, and Climate Volatility

Cities across the globe are relying mostly on infrastructure that was constructed for a different era, one with softer temperatures, slower urban expansion, and more consistent use patterns. With lengthy but limited lifespans, roads, bridges, tunnels, pipelines, electricity lines, dams, drainage systems, and transit networks were frequently built decades ago. Nowadays, a large number of these systems are getting older, breaking down, or performing below their designed limits. The strains on this aged urban infrastructure are becoming more severe as the climate issue worsens; they are no longer minor. Roads break under heat waves, drainage systems are overloaded by floods, concrete is cracked by freeze–thaw cycles (Mushtaq et al., 2024), and coastal fortifications are threatened by rising sea levels (Cheshmehzangi & Chen, 2021). The urban world's physical foundation is starting to crumble in too many locations due to mounting strain.

Infrastructure is the fundamental architecture of resilience, not just a group of technologies. It bears the burden of everyday existence by providing safe mobility, communication, power distribution, clean water, and shelter. These systems are frequently imperceptible when operating. The repercussions are severe and instantaneous when they fail. A bridge collapse, a water main break, or an extended blackout can cause cascading failures in emergency response, public health, and other areas (Zimmerman, 2004). And yet, maintenance frequently falls behind demand. Reactive repairs are made to infrastructure in many cities only after it has clearly deteriorated or failed. Tight funds, conflicting priorities, and bureaucratic slowness cause maintenance to be put off. Periodically occurring, labour-intensive, and frequently having a narrow scope are inspections. As a result, the gap between the operating demands placed on infrastructure and its physical state is growing.

This divide has proven perilous, as the climate emergency has shown. Twenty-first-century climate risks, such as longer droughts, more frequent hurricanes, heavier

© The Author(s), under exclusive license to Springer Nature Singapore Pte Ltd. 2025

A. Cheshmehzangi, *AI Agency in Climate Emergency*, Urban Sustainability,
https://doi.org/10.1007/978-981-95-4192-8_10

rainfall, and thermal extremes that hasten wear and stress, are posing a threat to infrastructure constructed in the previous century. What was formerly regarded as extraordinary has become commonplace, pushing the reliability of infrastructure to its limit. A new kind of urban fragility has been brought about by the combination of aged assets and climate volatility, in which vital systems fail due to inadequate monitoring, maintenance, or adaptation to present threats rather than faulty design.

This reality is beyond the capabilities of traditional maintenance approaches. They mostly rely on visual evaluations, planned inspections, and lagging failure signs. Unnoticed little fissures eventually develop into structural failures. Over the course of months or years, corrosion, wear, and water infiltration progress silently. A disruption, a collapse, or a disaster is frequently the initial indication of an issue. Deferred maintenance is extremely costly, not only monetarily but also in terms of lives lost, public confidence, and the social cohesion of urban areas. This model is no longer viable. Infrastructure in cities must be able to predict its own failure, indicate its own stress, and initiate its own repair.

Presenting AI as a novel form of urban guardian. AI presents a revolutionary approach to infrastructure management in this age of accumulating dangers, one that is based on alertness, anticipation, and ongoing feedback (Sepehr, 2025). A wide range of data sources, including embedded sensors, satellite imaging, LIDAR scans, drone footage, and digital twins of constructed objects, can be ingested by AI-enabled systems (Arun Sampaul Thomas et al., 2025). These systems can identify minute variations in temperature, vibration, load distribution, or material characteristics that indicate the beginning of deterioration. Cities can transition from reactive maintenance to predictive intervention with the use of machine learning algorithms, which can examine years' worth of performance data to find trends before failure (See Fig. 10.1).

AI holds the prospect of a new way of thinking, not only speed or automation. Infrastructure is reframed as an active system rather than a passive object, and as a speaking surface instead of a silent risk. Bridges that can monitor their own stress, water pipelines that can detect internal leaks in real time, and roads that can report their own cracking are examples of developing practices in cities that are taking resilience seriously. These technologies are no longer science fiction. AI enhances human control rather than replaces it, transforming sporadic inspections into ongoing monitoring and dispersed data into systemic awareness.

In this context, AI becomes a defender of public life as well as a technical advantage. It makes it possible for decision-makers to rank repairs according to risk rather than necessity. It guarantees that maintenance funds are allocated to the most critical areas. Additionally, it assists cities in facing the reality of climate adaptation before the next catastrophe strikes. AI's involvement will only increase as climate dangers increase and infrastructure ages. Although it is not a panacea, it is a potent instrument that helps us move away from a naive faith in steel and concrete and toward systems that are resilient, intelligent, and sensitive. Infrastructure is not merely constructed in the watchful city; it is observed, comprehended, and ready.

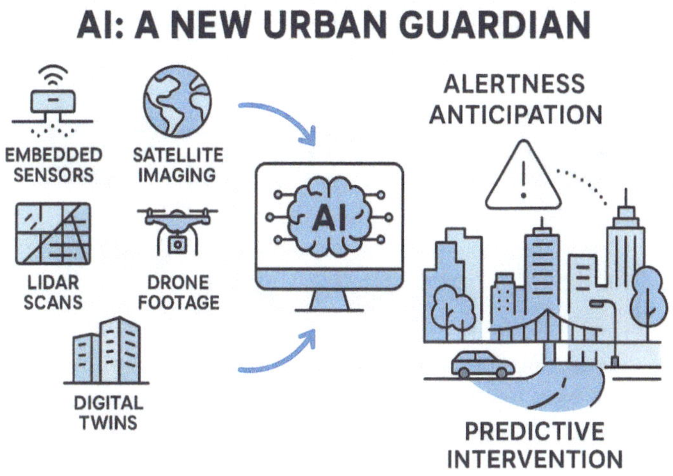

Fig. 10.1 This infographic illustrates AI as a vigilant urban guardian, integrating data from sensors, satellites, LIDAR, drones, and digital twins to enable predictive interventions—shifting cities from reactive maintenance to proactive infrastructure management through constant alertness and informed anticipation

10.2 Sensing and Predicting: How AI Monitors Structural Health

For a long time, urban administration has neglected the physical state of a city's infrastructure, including its pavements, bridges, tunnels, pipelines, and public buildings. Catastrophic failures garner media attention, but the quiet, plodding processes that lead to them are frequently overlooked until it's too late. These days, that reality is being drastically altered by the combination of AI and sensor technology. Our capacity to continuously, accurately, and in real time monitor the condition of infrastructure is being revolutionized by AI-integrated monitoring systems. Cities are given awareness and foresight by this intelligence, which enables them to take action before leaks turn into floods and fissures become collapses.

The implementation of dense sensor networks integrated into vital infrastructure is at the core of this change. Bridge vibration sensors, high-rise strain gauges, subterranean transport systems' temperature and humidity sensors, and water pipelines' acoustic sensors all produce real-time data on environmental conditions and structural behaviour. Little changes can be detected by these sensors, such as temperature spikes that show surface deformation in pavements, a slight change in resonance frequency that shows steel beam fatigue, or an increase in internal moisture levels that indicates water intrusion in concrete. These streams of data would be daunting on their own. However, AI turns them into trends, forecasts, and timely notifications.

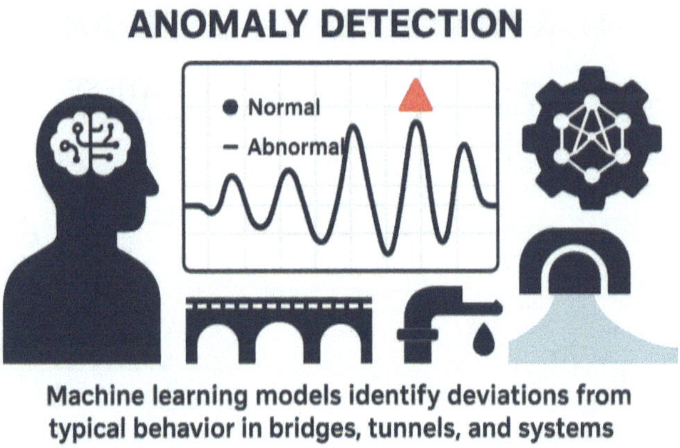

Fig. 10.2 This infographic demonstrates how machine learning models detect anomalies in infrastructure by distinguishing normal from abnormal behaviour. Using historical and real-time data, AI identifies early signs of deterioration in bridges, tunnels, and water systems, enabling proactive maintenance before visible damage occurs and preventing costly or dangerous failures

Anomaly detection—identifying departures from typical behaviour that can point to early-stage deterioration—is a specialty of machine learning models (Dasari et al., 2026). These algorithms, which are trained on historical data and real-time sensor inputs, identify what constitutes "normal" for a certain bridge, tunnel, or system and highlight any abnormalities that call for further examination. Even with years of experience, they are able to identify intricate, nonlinear patterns that human operators would probably overlook. For instance, long before eye examinations notice them, machine learning may identify minute alterations in structural resonance in bridges that point to growing fatigue or micro-fracturing. Even when the symptoms of subterranean leaks are not yet apparent at the surface, AI may detect changes in pressure signatures in water systems (See Fig. 10.2).

AI makes it possible to anticipate long-term infrastructure performance in addition to detection. AI systems may predict when and where materials are likely to degrade by combining material behaviour models with environmental data, such as rainfall patterns, humidity cycles, sun radiation, freeze–thaw events, and salinity exposure. These models anticipate stress accumulation in rail systems during heatwaves, estimate concrete expansion rates under intense heat, and simulate steel corrosion under various urban air quality scenarios. By predicting maintenance requirements and prolonging the useful life of vital infrastructure, such modelling assists cities in making the transition from reactive repair to strategic asset management.

By enhancing human vision, AI also improves infrastructure inspection. Drones fitted with heat sensors, LiDAR scanners, and high-resolution cameras have been used in recent years to visually assess assets that are difficult to reach, like the undersides of bridges, elevated roadways, and large-span roofs (Omar, 2018). These photos are processed by AI-powered computer vision, which provides remarkably accurate identification of surface fractures, spalling, corrosion areas, or material deformation. These systems have the ability to select regions for in-depth examination, map and quantify damage over time, and even generate condition reports automatically. They increase safety, lower labour expenses, and enable more frequent and thorough monitoring by decreasing the need for manual inspection.

These technologies' practical applications are already demonstrating their worth. In the US, states like California and Pennsylvania have benefited from AI-based bridge health monitoring initiatives funded by the Federal Highway Administration (Ebrahimpour et al., 2025). These technologies continuously evaluate the integrity of bridges and identify problems long before they endanger public safety by combining sensors and machine learning. AI is widely utilized to monitor bridge deformation and tunnel stress through distributed sensor arrays in Japan, where seismic risk exacerbates the requirement for resilient infrastructure. Following recent earthquakes, these technologies were crucial to the quick evaluation of damage and the prioritization of emergency responses by authorities (See Fig. 10.3).

Another striking example in the water sector is Singapore. The city-state has implemented a smart water grid that uses sensors and AI to track water pressure, flow rates, and quality throughout its whole pipeline system. Before extensive service interruptions happen, AI models can isolate trouble spots and identify anomalies that can indicate contamination or leaks. In a highly populated and water-constrained metropolitan setting, our proactive monitoring system guarantees system resilience and drastically lowers water loss.

These examples collectively demonstrate a new era of infrastructure intelligence, where prediction and sensing combine to build a city that is alert and responsive. AI transforms how we perceive, respond to, and plan for the changing demands of urban systems in a changing climate. It does more than just increase our ability to monitor infrastructure. The ability to listen to the city's physical systems and react quickly and intelligently may determine which cities prosper and which fail as infrastructure ages and hazards increase.

10.3 From Data to Action: Optimizing Maintenance and Lifecycle Resilience

It is no more an issue of whether or not to act, but rather how to act sensibly, effectively, and fairly as cities navigate an era characterized by both aging infrastructure and increasing climate stresses. Sensor-rich urban systems collect enormous amounts of data, but their full worth is only apparent when they are used to guide smart, timely decisions. In this equation, AI is becoming the key translator that turns raw data into actionable intelligence, assisting cities in identifying issues as well as prioritizing, organizing, and carrying out infrastructure repair with long-term resilience in mind.

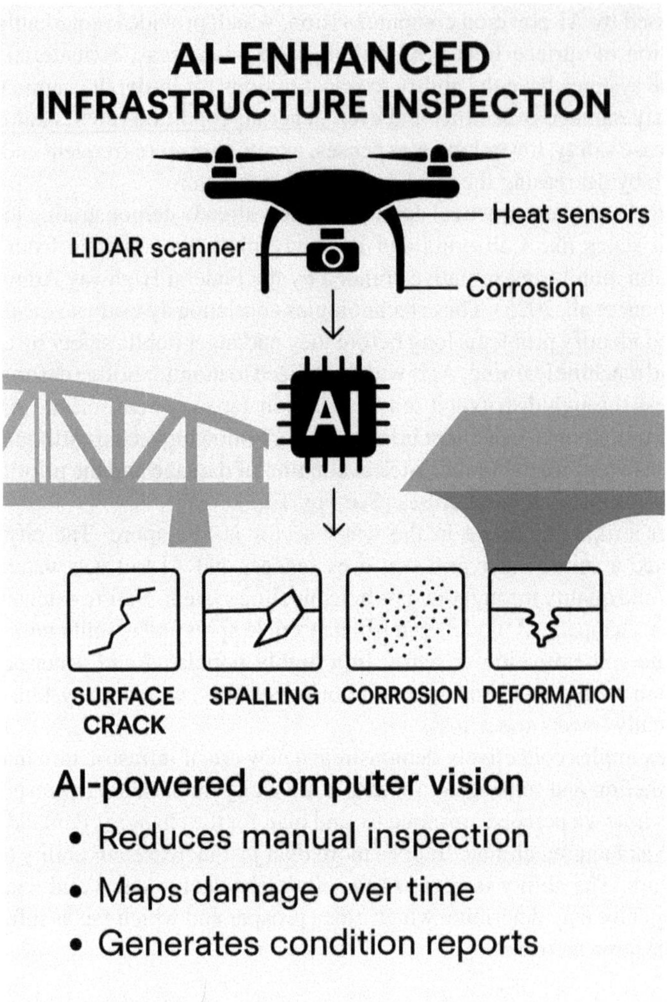

Fig. 10.3 AI-enhanced infrastructure inspection could use drones with LIDAR scanners and heat sensors. AI-powered computer vision then detects cracks, spalling, corrosion, and deformation, enabling safer, faster, and more accurate monitoring while reducing manual inspections and generating comprehensive condition reports over time

With this change, urban governance has undergone a fundamental transition from reactive patchwork to proactive systems thinking.

Prioritizing infrastructure renewal and repair tasks according to dynamic risk indicators is the first step in AI-driven maintenance optimization. AI models evaluate actual performance indicators, such as the volume and type of usage an asset experiences (Selvakumar et al., 2025), the frequency and severity of sensor-detected anomalies, and its exposure to environmental stressors like flooding, salt corrosion, or thermal cycling, instead of strict inspection schedules or financial quotas. For example, a regularly travelled road in a flood-prone location would require immediate reinforcing, whereas a bridge that is rarely used in a climate zone that is stable may safely postpone maintenance. This type of intelligent triage makes sure that limited resources—funding, labour, and materials—are distributed where they will have the biggest impact on resilience, safety, and service continuity.

AI facilitates comprehensive lifespan cost analysis and intervention planning in addition to triage. Machine learning algorithms can simulate the effects of delayed maintenance versus timely retrofitting by modelling the long-term behaviour of materials under various climate scenarios. This helps infrastructure managers make decisions based on projected performance and cost implications in addition to current condition. These technologies can predict how insulation systems would perform during extended heat waves, how shifting soil moisture will stress subterranean wires, or how a concrete retaining wall will deteriorate under frequent seawater flooding. Cities can use this information to make more informed judgments about whether to replace, repair, retrofit, or redesign—decisions that have an impact on finances and decades.

Importantly, AI also facilitates the incorporation of maintenance into more comprehensive resilience plans. More infrastructure is being modified for climate survival as well as functionality. Electrical substations can be made flood-proof, pavements can be coated with reflective materials to lessen heat absorption, rust-prone materials can be replaced with composites, or modular systems that can be improved without being completely replaced can be designed. By examining overlapping layers of structural vulnerability, urban density, and climate risk, AI can determine where such measures are most needed. For instance, it can suggest installing green roofs to reduce heat islands in heavily paved areas or raising power transformers in coastal areas where groundwater levels are rising. By coordinating maintenance with climate adaptation, cities can prepare for the extremes of the future rather than only fix the issues of the present.

Managing intricate, dispersed infrastructure networks demonstrates AI's capability even further. These infrastructures, which span thousands of kilometers and frequently lay beneath or behind visible city activity, include roads, water mains, power lines, and stormwater drains. AI-powered asset management systems keep an up-to-date, dynamic map of system health by continuously ingesting data from sensors, inspections, satellite imaging, and historical documents. These platforms can recognize patterns across systems, such as the fact that power line problems tend to cluster in locations with higher tree density and wind exposure, or that specific road segments deteriorate more quickly according to the makeup of the traffic. With the use of these data, communities may transition from patchwork maintenance to comprehensive network optimization, reducing downtime, increasing longevity, and enhancing service delivery equity.

Examples from around the world provide strong support for this change. AI is used to predict material wear and surface degradation in Scandinavian countries, where freeze–thaw cycles severely damage roads and bridges. To plan preventative maintenance before cracking and potholes appear, these models take into consideration temperature variations, snow load, ice expansion, and de-icing agents. To prevent expensive emergency interventions in the winter, municipalities can then effectively coordinate repairs during the short summer windows. Cities like Abu Dhabi and Doha are utilizing AI to prioritize heat-resilient retrofits, such as upgrading to UV-resistant cladding, insulating subterranean cabling, and strengthening stormwater systems to withstand severe flash flooding, in the Persian Gulf region, where extreme heat, sandstorms, and salty air cause infrastructure to corrode quickly.

In both situations, AI provides a new logic of infrastructure stewardship in addition to engineering precision. This reasoning acknowledges that operational effectiveness and climate resilience are complementary rather than antagonistic objectives. It prioritizes intelligence over inertia, systems over silos, and foresight over firefighting. Every asset in the watchful city turns into a sensor, every fix into a plan, and every maintenance schedule into an opportunity to prolong not just the lifespan of the infrastructure but also the city's overall liveability.

10.4 Governance, Ethics, and Systemic Resilience

The deployment of AI-powered infrastructure technologies, which are becoming increasingly essential to urban resilience, requires careful governance, moral foresight, and institutional coordination (Cheshmehzangi, 2025). Although the technology capabilities for adaptive response, real-time monitoring, and predictive maintenance have advanced quickly, the frameworks for their equitable and efficient deployment frequently lag behind. The adoption of smart infrastructure solutions is still being slowed by issues with interagency cooperation, regulatory approval, and public procurement, especially in urban areas where outdated bureaucracy and legacy systems stifle innovation. Intelligent infrastructure deployment is a systems change that necessitates new forms of accountability, transparency, and governance; it is not just a technical improvement.

The inflexibility of public procurement procedures is one of the most pressing obstacles. Governments are frequently constrained by antiquated bidding procedures that prioritize the lowest price over long-term worth or are unable to adapt to rapidly changing technologies like AI. Additionally, regulatory agencies might not be ready to evaluate and authorize AI-powered inspection tools or predictive maintenance systems, particularly if the underlying algorithms are private or opaque. Reforming procurement procedures as well as how cities assess, and license new digital tools is necessary to remove these obstacles. Every AI deployment needs to be supported by trust, openness, and thorough assessment, especially in systems that have an impact on public safety and urban operations.

The core of infrastructure surveillance must also be equity. Under the pretence of safety or resilience, there is a genuine risk that some communities—typically lower-income or historically marginalized—will be over-monitored as cities integrate sensors and AI tools into public areas. Existing surveillance disparities must not be repeated or exacerbated by infrastructure intelligence. To ensure that data gathering focuses on physical systems (Gouglidis et al., 2018) rather than people, cities must establish explicit criteria that differentiate between asset monitoring and community surveillance. Communities should also be actively involved in deciding who will have access to the data produced by monitoring technologies, as well as how and where they will be implemented.

The importance of human expertise is also called into doubt by the growing automation. Even if AI is very good at pattern recognition, predictive modelling, and anomaly detection, engineers, inspectors, and infrastructure managers still need their situational awareness, moral discernment, and field experience (Popoola & Bakare, 2024). Augmented decision-making, where AI complements human understanding rather than replaces it, is the way of the future. AI-generated forecasts and diagnostics must continue to assist the human-led process of maintenance planning. This equilibrium is essential for maintaining professional knowledge and civic stewardship as well as for trust and responsibility.

The increasing trend toward collaborative governance and open-source infrastructure intelligence provides a positive contrast to these difficulties (Buckingham Shum et al., 2012). The possibility for collaboration increases when data, models, and monitoring tools are exchanged and made transparent among agencies. Road repairs and underground utility upgrades can be coordinated by transportation bureaus. Forecasts of drainage capacity associated with building development are available to flood resilience teams. Civic tech groups can create local dashboards to raise awareness of climate risk or crowdsource maintenance reports using open datasets. Disjointed city departments may become adaptable, learning ecosystems through this systemic integration, which is driven by open standards, shared platforms, and AI-enabled analysis.

Finally, AI in infrastructure holds the potential of foresight as much as efficiency. It enables cities to become proactive instead of reactive, to step in before there is a disturbance, and to allocate scarce resources in a way that maximizes the good of the public (Moraci et al., 2018). This intelligence is essential in a time of escalating hazards and constrained urban finances. The ethical city must be strategic without compromising inclusivity, efficient without deleting human judgment, and watchful without being intrusive. AI can help cities not just survive climatic stress but also thrive by enabling them to adapt and change before systems collapse, provided it is applied carefully and strategically.

References

Buckingham Shum, S., Aberer, K., Schmidt, A., Bishop, S., Lukowicz, P., Anderson, S., Charalabidis, Y., Domingue, J., de Freitas, S., Dunwell, I., & Edmonds, B. (2012). Towards a global participatory platform: Democratising open data, complexity science and collective intelligence. *The European Physical Journal Special Topics, 214*(1), 109–152.

Cheshmehzangi, A., & Chen, H. (2021). Cities and climate-resilient development. In *China's sustainability transitions: Low carbon and climate-resilient plan for carbon neutral 2060* (pp. 11–19). Springer Singapore.

Cheshmehzangi, A. (2025). Bridging Innovation and Sustainability: Global Insights on ICT Applications. In *Urban innovation and sustainability: ICT-based interventions in Hainan, China* (pp. 23–43). Springer Nature Singapore.

Dasari, K., Parry, S., & Mekala, S. (2026). Transforming aerospace design and manufacturing through machine learning. In *Innovative machine learning applications in the aerospace industry* (pp. 57–86). IGI Global Scientific Publishing.

Ebrahimpour, A., Baibordy, A., & Ibrahim, A. (2025). Developing a research roadmap for highway bridge infrastructure innovation: A case study. *Infrastructures, 10*(6), 133.

Gouglidis, A., Green, B., Hutchison, D., Alshawish, A., & de Meer, H. (2018). Surveillance and security: Protecting electricity utilities and other critical infrastructures. *Energy Informatics, 1*(1), 15.

Moraci, F., Errigo, M. F., Fazia, C., Burgio, G., & Foresta, S. (2018). Making less vulnerable cities: Resilience as a new paradigm of smart planning. *Sustainability, 10*(3), 755.

Mushtaq, M., Corradi, M., & Sikdar, S. (2024). Climate adaptation of roads to flooding hazards—A review. *Construction Materials (2673–7108), 4*(4).

Omar, T. (2018). *Condition assessment of concrete bridge decks using ground and airborne infrared thermography* (Doctoral dissertation, The University of Western Ontario (Canada)).

Popoola, N. T., & Bakare, F. A. (2024). Advanced computational forecasting techniques to strengthen risk prediction, pattern recognition, and compliance strategies. *International Journal of Science and Research Archive, 12*(02), 3033–3054. https://doi.org/10.30574/ijsra.2024.12.2.1412

Selvakumar, P., Murugan, S., Gupta, P. K., Reddy, D. S., & Srivastava, P. (2025). Employment of Ml and AI in asset management and evaluation. In *Utilizing AI and machine learning in financial analysis* (pp. 81–100). IGI Global Scientific Publishing.

Sepehr, P. (2025). Mundane urban governance and AI oversight: The case of Vienna's intelligent pedestrian traffic lights. *Journal of Urban Technology, 32*(3), 93–110.

Arun Sampaul Thomas, G., Muthukaruppasamy, S., Sathish Kumar, S., Karthikeyan, B. J., & Saravanan, K. (2025). Navigating the Nexus: Unravelling Challenges, Ethics, and Applications of Embodied AI in Drone Technology Through the Lens of Computer Vision. In *Building embodied ai systems: the agents, the architecture principles, challenges, and application domains* (pp. 61–78). Springer Nature Switzerland.

Zimmerman, R. (2004). The epidemiology of extreme events: Health and safety impacts of cascading infrastructure failures. *Epidemiology, 15*(4), S126–S127.

Part IV
AI for Climate-Smart Land Use and Agriculture

To fight climate change, this section examines how AI is promoting sustainable land management and agricultural methods. AI is being used in precision agriculture to enhance pest management, fertilization, and irrigation while lowering emissions and resource consumption. To improve the monitoring of natural carbon sinks and aid in climate mitigation, machine learning algorithms measure soil carbon and forest growth. In the meantime, near real-time identification of land degradation and deforestation is made possible by AI-powered analysis of satellite imagery, offering vital tools for ecosystem protection. The effective, climate-resilient, and ecologically conscious use of land and agricultural systems is facilitated by these developments.

Chapter 11
Farming with Foresight: AI for Precision Agriculture and Resource Efficiency

11.1 From Uniform to Precise: Rethinking Agriculture for a Warming Planet

The principle of uniformity has governed agriculture for a large portion of modern history. Regardless of minute variations in soil quality, moisture content, or plant health, large fields were handled as single, undifferentiated units that were planted, watered, fertilized, and sprayed in identical proportion. This input-heavy strategy, which was intended for simplicity and scale, increased yields at a high cost: increased greenhouse gas emissions, soil exhaustion, water waste, chemical runoff, and biodiversity loss. Even as profits per input started to level out, agriculture's environmental impact increased as it spread into more marginal territories. This homogeneous paradigm has reached its limits in the context of a warming globe.

The environment that farmers face today is drastically different, characterized by ecological fragility, climate instability, and the pressing need to decarbonize. Rainfall is becoming more unpredictable, pest outbreaks are changing in scope and intensity, and droughts are becoming more prolonged and severe. There is now widespread soil degradation, which is frequently made worse by excessive fertilizer use and mechanical tillage. Much of the Global South and even some areas of the historically fertile Global North are plagued by water scarcity. However, when land use change and food systems are taken into account, agriculture continues to be a major source of emissions, contributing about 25% of global greenhouse gas emissions. The business-as-usual paradigm is no longer viable due to contemporary agriculture's dual goal of feeding a growing population while lowering its environmental impact.

A new logic is developing in this situation, one that prioritizes accuracy over consistency. By seeing every square meter of farmland, or even every single plant, as a distinct data point with distinct demands and reactions, precision agriculture signifies a fundamental change away from broad-scale interventions and toward hyper-local decision-making. This method optimizes water use, nutrient application, insect management, and harvesting schedules by utilizing site-specific information.

© The Author(s), under exclusive license to Springer Nature Singapore Pte Ltd. 2025

A. Cheshmehzangi, *AI Agency in Climate Emergency*, Urban Sustainability,
https://doi.org/10.1007/978-981-95-4192-8_11

Farmers act on evidence and variability rather than on averages and assumptions. This change offers a practical route toward climate adaptation and mitigation in agriculture, in addition to increasing efficiency.

The core of this shift is AI and its extended applications. AI makes farming adaptive, predictive, and self-correcting in addition to digitizing it (Jayaraman Krishnamurthy, 2023). AI systems generate a comprehensive picture of farm conditions over time and space by combining satellite photos, drone surveillance, soil sensors, weather forecasts, and real-time field data. By mapping soil nutrient zones, predicting how a particular hybrid crop will react to a late-season heatwave, or recognizing disease start before symptoms appear, machine learning models may assess this variability and reveal patterns that are unseen to the human eye. Precise interventions are subsequently guided by these insights, such as modifying irrigation zones according to soil moisture, applying fertilizer at times that correspond with plant absorption peaks and rainfall events, or focusing pest hotspots with minimal pesticide use.

The ability of AI to continuously learn and adapt is what gives it power in this situation. A dynamic algorithm—one that changes with every season, field scan, and feedback loop—can adapt to the shifting climate of today better than a static guideline (Assilian, 1974; Eddamiri et al., 2025). AI systems can help farmers get ready for drier years or changing frost dates by simulating future conditions based on climate models. Instead of using regional averages, they can optimize planting dates depending on localized microclimates. Additionally, they can contribute to lowering emissions by reducing the excessive use of nitrogen, eliminating needless tillage, or better controlling fuel and energy consumption using automated robotics and mechanical routes.

Importantly, AI-powered precision agriculture aims to create climate-resilient food systems from the ground up, not only increase yields or reduce expenses (Patil, 2024). It makes it possible for farming that minimizes emissions, preserves soil health, conserves water, and instantly adjusts to environmental stressors (Thingujam et al., 2025). This makes it a mitigation technique that reduces agriculture's carbon footprint while maintaining ecosystem integrity, as well as an adaptation method that helps farmers deal with unstable situations.

The basic foundations of agriculture are being forced to be rethought in light of the climate catastrophe. Once considered a luxury or high-tech accessory, precision is quickly becoming essential. This new paradigm, powered by AI, promises a more secure and sustainable future for food production on a warming globe in addition to smarter farms.

11.2 Targeted Irrigation: AI for Water Efficiency and Drought Resilience

One of the most limited and valuable resources in agriculture has always been water. For farms all around the world, intelligent and efficient water usage has become essential to survival in the face of increasing climate pressure, which includes longer droughts, unpredictable rainfall, and depleting aquifers (Ahmed et al., 2023). Overwatering, runoff, or stress from under-irrigation are common outcomes of traditional irrigation techniques, which are typically based on set timetables or intuition. These inefficiencies are unsustainable in addition to being expensive. Approximately 70% of freshwater withdrawals worldwide are currently used for agriculture, and as water shortage worsens, there is an urgent need for more intelligent, flexible irrigation. Here comes AI, a potent ally in converting water consumption from reactive to predictive and from approximation to precise.

The combination of machine learning and Internet of Things (IoT) technologies—more especially, soil moisture sensors, climate monitoring stations, and remote sensing systems—is the foundation of AI-enabled irrigation. Real-time data on soil saturation, plant water requirements, local humidity, temperature, and rainfall projections are produced by these instruments. Together with evapotranspiration models, AI systems evaluate this data to determine how much water is lost from soil and plant surfaces as a result of heat, wind, and sunlight. By preventing waste and stress, this integration enables dynamic, field-level insights about the amount, location, and timing of water requirements.

Precision zoning and scheduling are made possible by AI-guided technologies, as opposed to distributing water evenly throughout a field (Arogundade & Njoku, 2024). By identifying minute differences in crop type, geography, and soil composition, these technologies enable farmers to water only the regions that require it—and only to the necessary degree. Compared to a shady, loamy area in the same field, a section of land on a slope with sandy soil could require shorter, more frequent watering (Ullah et al., 2009). AI regularly modifies these schedules in response to weather variations, anticipating decreased irrigation requirements in the event of an impending rainstorm or adapting to a sudden drop in humidity. Real-time irrigation customization guarantees that plants get the right amount of water, increasing production and resource efficiency (See Fig. 11.1).

AI also improves sophisticated irrigation methods like deficit and drip irrigation. AI adjusts flow rates and timing in drip systems, which use a network of emitters to distribute water straight to plant roots, to match the crop's precise absorption capability (Fang, 2023). This enhances plant health, lowers nutrient leaching, and conserves water. AI tracks plant stress levels and environmental circumstances to administer minimal but sufficient irrigation without sacrificing production in deficit irrigation, which purposefully reduces water during specific growth phases to cause stress responses and conserve resources. AI can identify early indicators of water stress that are not visible to the human eye and take corrective action before yields

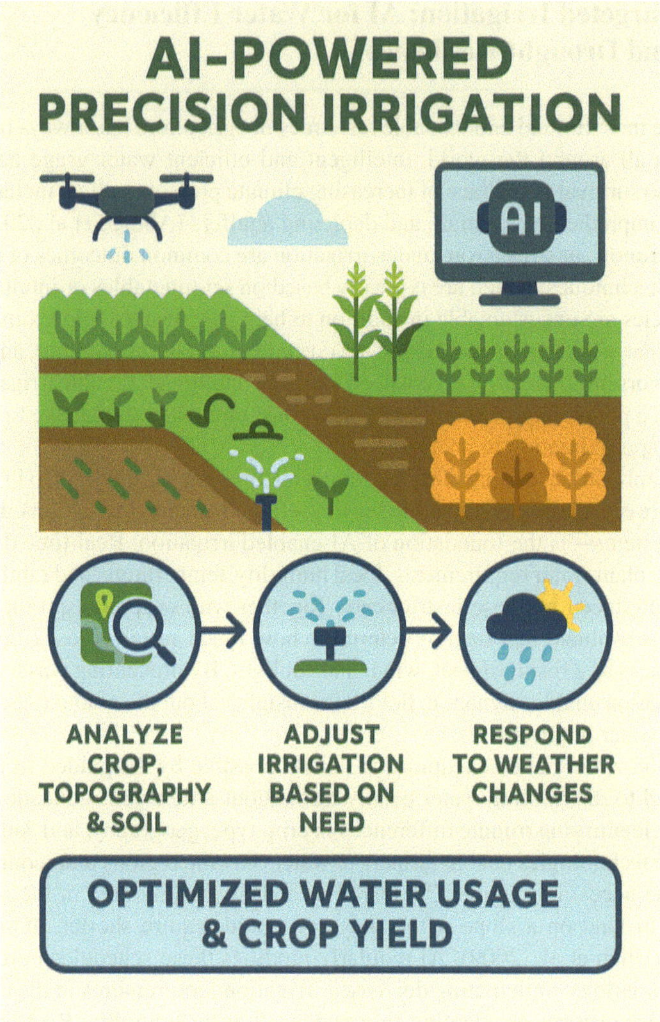

Fig. 11.1 AI-powered precision irrigation, where drones, sensors, and AI analysis optimize water use by analyzing crop type, topography, and soil. Irrigation schedules adjust dynamically to specific needs and weather changes, ensuring efficient water distribution, higher crop yields, and reduced waste across diverse agricultural zones

are impacted by using thermal imaging, multispectral drone surveys, and canopy temperature analysis.

These systems' practical uses are demonstrating their value. AI-integrated irrigation networks are commonplace in dry regions of China, a world leader in agricultural water efficiency (Cheshmehzangi et al., 2021). Real-time soil and crop data is used by systems such as Netafim's precision irrigation platforms to automatically regulate watering schedules, saving a lot of water and increasing crop yield. AI-powered solutions like Tule and CropX, which integrate satellite imaging, field sensors, and machine learning to inform irrigation strategies down to the individual row, are being used by vineyards and almond orchards in California's Central Valley, where drought has become a structural concern. Government-backed projects in India are introducing AI-driven smartphone apps and automated irrigation pumps that respond to rainfall patterns and moisture levels, lowering water consumption and enhancing resilience in areas with historically unstable irrigation access. Smallholder farmers in India are particularly susceptible to monsoon variability (See Fig. 11.2).

These innovations have a significant influence. Crops are grown in more stable, healthier circumstances, water waste is significantly decreased, and groundwater depletion is slowed. Even in years that are prone to drought, farmers enjoy greater yields, save money on power, and require less emergency watering. Most significantly, farmers can act with confidence rather than relying on conjecture thanks to real-time, data-informed intelligence, which lessens the psychological load of ambiguity.

AI-powered irrigation is unquestionably beneficial for production and sustainability. When handled carefully and precisely, it transforms water into a strategic asset. Such foresight is essential to the future of climate-resilient agriculture in a warming world when every drop matters.

11.3 Smarter Inputs: Fertilization, Pest Control, and Emissions Reduction

Fertilizers, insecticides, and herbicides are intensive inputs that have long been used in modern agriculture to increase yields and manage risks. However, the uniform and frequently excessive application of these inputs have a negative impact on the environment, leading to large greenhouse gas emissions, biodiversity loss, water contamination, and soil deterioration. Although nitrogen fertilizers are necessary for agricultural growth, they also release nitrous oxide, a greenhouse gas that is almost 300 times more powerful than carbon dioxide. On the other hand, overuse of pesticides causes chemical resistance in pests and endangers pollinator populations. Traditional blanket-input methods are proven to be inefficient and wasteful as climate change changes fertilizer cycling, rainfall timing, and pest migration. Data-driven and AI-powered, a more shrewd, more precise approach is required.

By offering real-time, site-specific fertilization and pest control recommendations, AI puts intelligence at the core of input management. Machine learning algorithms use nutrient mapping to create variable-rate fertilization plans by analyzing crop type, weather forecasts, soil composition, and historical yield data. By lowering over-application in nutrient-rich zones and increasing it where shortfalls occur, these programs modify nutrient application not just between fields but also within a single plot. AI-guided fertilization increases yields while minimizing runoff into rivers and

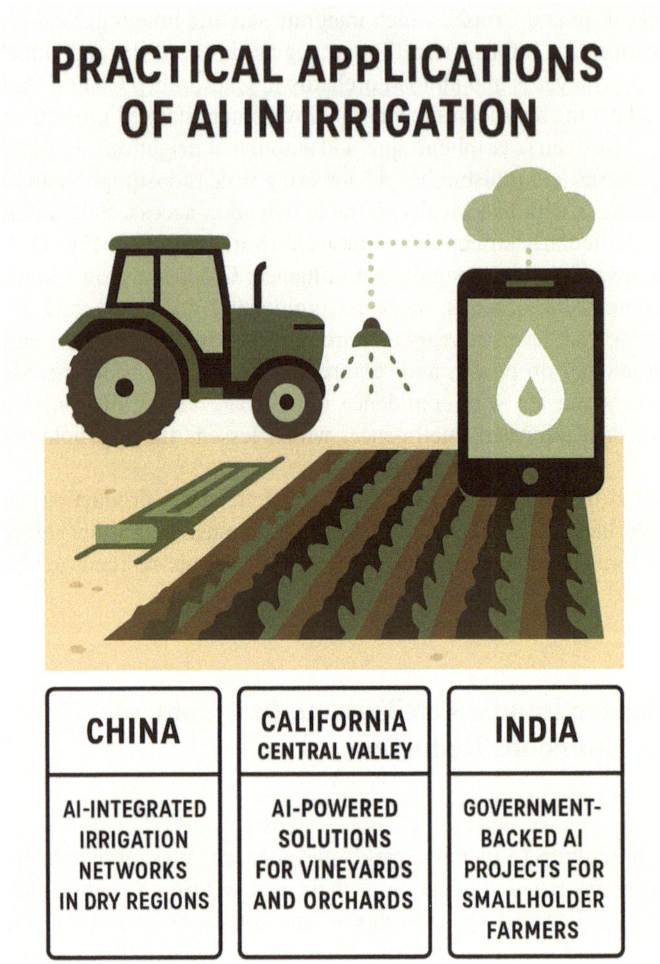

Fig. 11.2 This infographic shows practical AI applications in irrigation: China's AI-integrated networks boost water efficiency in dry regions, California's AI-powered vineyard and orchard systems optimize drought resilience, and India's government-backed AI projects support smallholder farmers with smart irrigation responsive to rainfall and soil conditions

lowering the possibility of nitrogen loss into the atmosphere by perfectly balancing nutrient delivery with plant demand (see Fig. 11.3).

Additionally, predictive modelling is essential for managing diseases and pests. Based on particular environmental triggers, AI systems trained on entomological databases, plant pathology records, and climatic patterns can predict the probability of pest infestations or disease outbreaks. A sudden increase in temperature and humidity, for instance, can indicate that the conditions are right for the development of an insect

Fig. 11.3 Showcasing AI-driven input management, where machine learning analyzes crop type, weather forecasts, soil composition, and historical yields to deliver variable-rate fertilization. By tailoring nutrient application within fields, AI boosts crop yields, reduces runoff, and minimizes nitrogen loss, aligning nutrient supply precisely with plant demand for sustainable farming

life cycle in rice paddies or for fungal growth on grapevines. In order to give farmers early warnings before issues spread, these models are continuously updated using satellite and drone footage, plant health indices (such the Normalized Difference Vegetation Index, or NDVI), and in-field sensor readings. Growers can precisely apply preventative remedies instead of responding after the fact.

The usage of autonomous robots and drones directed by AI further improves this accuracy. These computer vision-equipped systems, which have been trained on thousands of crop photos, are able to recognize weed clusters, disease symptoms, and insect presence at the plant level. After that, they carry out ultra-targeted spraying, which is the process of applying tiny amounts of pesticide or herbicide just where necessary and frequently with extreme precision. Drones are able to selectively spray antifungals and detect powdery mildew in grapes. Robotic sprayers that can discriminate between healthy crops and sick areas can drastically cut down on the amount of chemicals used in rice and wheat fields (Meshram et al., 2022). This significantly reduces the environmental impact of farming activities while also saving money and resources.

Emissions reduction is one of the most important effects of this focused strategy. Farmers can avoid excess nitrogen turning into nitrous oxide by optimizing nitrogen application with AI-informed scheduling and dose. When AI is used in conjunction with weather forecasting and soil moisture monitoring, it guarantees that fertilizer is delivered when plants are most able to absorb it. By lowering volatilization, leaching, and gaseous emissions, this helps achieve climate mitigation objectives as well as farm profitability. Additionally, by preserving beneficial insects and soil organisms that are essential to the long-term health of the soil, the decrease in pesticide use contributes to the restoration of ecological balance.

Numerous crops and geographical areas are already utilizing these concepts. In order to minimize the number of fungicide sprays required per season while preserving high-quality yields, grape producers in France and Spain are employing AI-driven disease forecasting models to schedule their fungicide applications. AI-driven nutrient management systems are being used in rice fields in Southeast Asia, where variable-rate fertilizer delivery is assisting smallholder farmers in increasing yield while reducing expenses. AI systems based on soil sampling and remote sensing are helping farmers in Pakistan and India produce wheat by advising them on when and how much urea to use, which lowers runoff into rivers and improves grain quality.

The transition from wide input utilization to hyper-specific, ecologically informed decision-making is exemplified by these success stories. AI is assisting agriculture in its shift from over-treatment to precision—from general care to focused attention. It aligns farm-level decisions with global sustainability goals and promotes environmental responsibility in addition to operational efficiency. By doing this, AI transforms from a tool for productivity into one that promotes agricultural regeneration and helps bring yield, economy, and environment back into balance in a warming planet.

11.4 Scaling Sustainability: Equity, Access, and Climate-Smart Transitions

The realization that the advantages of AI-powered precision agriculture must be shared fairly is growing along with the technology's potential. The situation for millions of smallholder farmers, especially in the Global South, is drastically different from that of large-scale commercial farms in technologically sophisticated nations, which are already benefiting from AI-driven crop analytics, autonomous sprayers, and smart irrigation systems. Adoption is nevertheless hampered by issues such high upfront prices, inadequate internet connectivity, low levels of digital literacy, and restricted access to customized data services. Precision agriculture runs the potential of escalating rather than reducing the global disparity in food security, climate resilience, and agricultural output if these issues are not resolved.

Technology is insufficient on its own. The true test will be in creating support systems that make AI tools available, practical, and advantageous for all agricultural sizes. This shift is already being significantly aided by open-source platforms. These efforts encourage local customization and reduce entry costs by making AI algorithms, agricultural models, and climate forecasts publicly available. Examples of how digital infrastructure can empower farmers, agronomists, and cooperatives in areas where commercial platforms may be unavailable include FAO's WaPOR project, which offers open-access satellite-based evapotranspiration data (Barideh & Nasimi, 2022), and PlantVillage (Mrisho et al., 2020), which uses AI to diagnose plant diseases via smartphones.

Another important enabler is government support. AI adoption can be significantly accelerated by policy frameworks that promote investment in agricultural technology, provide smallholders with subsidies for precise equipment, and increase internet connectivity in rural areas. For example, AI chatbots in regional languages have been used by public–private efforts in India to provide insect warnings and planting guidance. Together with mobile-based advising systems designed to meet the needs of smallholders, governments and non-governmental organizations in East Africa have teamed up to distribute solar-powered IoT sensors for irrigation control. These initiatives not only boost output but also integrate digital agriculture with more general objectives of climate resilience and rural development.

Farmer networks and agricultural cooperatives are also essential. Cooperatives may democratize access to AI-powered tools and training by combining infrastructure, resources, and expertise. Drones, field sensors, and data analytics platforms can be owned jointly, which helps lower expenses and foster group decision-making. Additionally, cooperatives act as reliable middlemen, bridging the gap between state-of-the-art technology and practical realities. AI becomes a bottom-up enabler, rooted in farmers' lived experiences and their land, when combined with local knowledge and community governance.

AI plays a significant role in enhancing farmer resilience through timely information, going beyond tools and platforms. Early warnings about weather extremes, insect outbreaks, and market swings can be provided via intelligent alert systems that are distributed via SMS, apps, or radio. By predicting price trends, predictive analytics can assist farmers in selecting more resilient crop varieties, precisely scheduling the planting and harvesting processes, and negotiating better terms in supply chains. When ethically and inclusively built, these systems give farmers a strategic advantage in negotiating the risks of climate-driven agriculture in addition to technical guidance.

Ultimately, AI in agriculture aims to develop resilience rather than just better crops. Its goals are to assist farmers in protecting their ecosystems, producing more with less, and adjusting to the ever-changing natural landscape (Shafik et al., 2025). Precision agriculture provides a scientifically supported and socially responsive route to sustainability in a world where food systems are being strained by climatic extremes, geopolitical unrest, and growing population demands. However, this route needs to be inclusive.

AI-powered agriculture has to develop into a future shared infrastructure rather than merely a business venture (Cheshmehzangi, 2025). It needs to be fostered through international solidarity, cooperative governance, and public investment. Only then can its promise be fulfilled—not as a luxury for a select few, but as a foundation for climate-smart changes that increase yields, restore land, and ensure food security in the face of an uncertain future. According to this perspective, AI is not just a digital invention but also a vital component of the global jigsaw for sustainable development.

References

Ahmed, Z., Gui, D., Murtaza, G., Yunfei, L., & Ali, S. (2023). An overview of smart irrigation management for improving water productivity under climate change in drylands. *Agronomy, 13*(8), 2113.

Assilian, S. (1974). *Artificial intelligence in control of real dynamic systems* (Doctoral dissertation). Queen Mary University of London.

Arogundade, J. B., & Njoku, T. K. (2024). Maximizing crop yields through AI-driven precision agriculture and machine learning. *International Research Journal of Modernization in Engineering Technology and Science, 6*(10), 1022–1042, https://doi.org/10.56726/IRJMETS62193

Barideh, R., & Nasimi, F. (2022). Investigating the changes in agricultural land use and actual evapotranspiration of the Urmia Lake basin based on FAO's WaPOR database. *Agricultural Water Management, 264*, Article 107509.

Cheshmehzangi, A. (2025). Bridging innovation and sustainability: Global insights on ICT applications. In *Urban innovation and sustainability: ICT-based interventions in Hainan, China* (pp. 23–43). Springer Nature Singapore.

Cheshmehzangi, A., Dawodu, A., & Sharifi, A. (2021). *Sustainable urbanism in China*. Routledge.

Eddamiri, S., Bassine, F. Z., Hakam, O., Kambiet, P. L. K., & Chehbouni, A. (2025). Applications of artificial intelligence for forecasting and managing extremes. In *Climate change and rainfall extremes in Africa* (pp. 91–113). Elsevier.

Fang, H. (2023). Actionable science for irrigation. *Actionable science of global environment change: From big data to practical research* (pp. 203–228). Springer International Publishing.

Jayaraman Krishnamurthy, R. (2023). *Integration of an AI-based self-correcting fused deposition modelling, composite sensor, and XR technology: An Industry 4.0 demonstration study* (Doctoral dissertation, University of British Columbia).

Meshram, A. T., Vanalkar, A. V., Kalambe, K. B., & Badar, A. M. (2022). Pesticide spraying robot for precision agriculture: A categorical literature review and future trends. *Journal of Field Robotics, 39*(2), 153–171.

Mrisho, L., Mbilinyi, N., Ndalahwa, M., Ramcharan, A., Kehs, A., McCloskey, P., Murithi, H., Hughes, D. & Legg, J. (2020). Evaluation of the accuracy of a smartphone-based artificial intelligence system, PlantVillage Nuru, in diagnosing of the viral diseases of cassava. *BioRxiv*, 2020-01.

Patil, D. (2024). *Artificial intelligence innovations in precision farming: Enhancing climate-resilient crop management.* Available at SSRN 5057424.

Shafik, W., Tufail, A., De Silva, C. L., Haji, R. A. A., & Apong, M. (2025). The role, application, and impact of artificial intelligence in the agriculture industry. In *Future Tech Startups and Innovation in the Age of AI* (pp. 36–60). Crc Press.

Thingujam, D., Gouli, S., Cooray, S. P., Chandran, K. B., Givens, S. B., Gandhimeyyan, R. V., Tan, Z., Wang, Y., Patam, K., Greer, S. A., & Acharya, R. (2025). Climate-resilient crops: Integrating AI, multi-omics, and advanced phenotyping to address global agricultural and societal challenges. *Plants, 14*(17), 2699.

Ullah, R., Lone, M. I., Ali, S., & Hussain, S. (2009). Soil water variation under different cropping patterns on sloppy lands in Punjab, Pakistan. *Soil & Environment, 28*(2), 156–161.

Chapter 12
Carbon Sequestration Tracking: AI for Measuring the Invisible in Soil, Forests, and Landscapes

12.1 Nature as a Carbon Sink: The Promise and Complexity of Land-Based Sequestration

Nature's capacity to store carbon makes it one of the most effective—yet underutilized—tools in the fight against climate change. Through photosynthesis, soils, forests, wetlands, and grasslands absorb atmospheric CO_2 and store it in biomass and subsurface organic matter, making them enormous, living carbon sinks. Soils store three times as much carbon as the atmosphere, and forests store over a quarter of the world's carbon. In addition to sequestering carbon, healthy ecosystems maintain livelihoods, control water cycles, promote biodiversity, and offer resilience against climate shocks (Daba & Dejene, 2018). The promise of nature-based solutions (NbS) has become a focal point of global mitigation initiatives as the climate catastrophe worsens (Cheshmehzangi, 2022; Cheshmehzangi et al., 2018; Li et al., 2023; Seddon et al., 2020). Yet, to fully realize this potential, we must first learn how to see and measure the invisible processes of sequestration at the scale and precision required.

The concept of employing natural systems as climate infrastructure has gained popularity throughout the last ten years. To improve carbon sinks, governments, corporations, and conservation organizations are funding regenerative agriculture, wetland restoration, rewilding, and tree planting projects. These land-based sequestration initiatives are now included in business sustainability commitments, carbon offset markets, and national climate strategies (Barber et al., 2024). But as these initiatives grow, so does the need for timely, accurate, and transparent carbon accounting. What is the actual amount of carbon sequestered? How long? And in what circumstances? One of the most enduring problems facing climate action today is finding answers to these issues.

The main challenge is measurement. Conventional techniques for measuring soil and biomass carbon include field surveys, laboratory analysis, and manual sampling; these methods are costly, time-consuming, and geographically restricted (Loria et al., 2024). Field workers are required to measure tree diameters, calculate canopy density,

© The Author(s), under exclusive license to Springer Nature Singapore Pte Ltd. 2025
A. Cheshmehzangi, *AI Agency in Climate Emergency*, Urban Sustainability,
https://doi.org/10.1007/978-981-95-4192-8_12

and infer carbon content from statistical models in forest carbon projects. Soil carbon is usually tested across small plots and at shallow depths in agriculture, which ignores heterogeneity across depths and regions. Despite being scientifically sound, these approaches are unable to satisfy the requirements of global scalability or real-time verification. The heterogeneity of natural systems—how carbon stocks differ among topographies, seasons, soil types, and land-use practices (Fernández-Romero et al., 2014)—is another aspect of natural systems that they find difficult to depict.

AI becomes a transformative force in this situation. AI makes it possible to track, model, and forecast carbon sequestration dynamically and at scale, particularly when paired with remote sensing technology. Large volumes of geospatial data, including those from satellites, drones, field sensors, and historical documents, can be ingested by machine learning algorithms, which can then identify patterns that human analysts would be unable to identify. These models can forecast carbon accumulation trajectories under various land management scenarios, estimate biomass from canopy reflectance, and predict soil organic carbon based on spectral signatures.

AI also aids in overcoming fragmented and low-resolution data, which is one of the biggest obstacles to nature-based carbon tracking. Consistent carbon inventories and monitoring systems are lacking in many places, particularly in the Global South (Twyman et al., 2015). Even sparse datasets can be enhanced and interpolated with AI through ensemble modelling and transfer learning. For example, models for neighbouring regions with only moderate-resolution satellite data can be calibrated using forest carbon estimates from high-resolution lidar data in one location. In order to create comprehensive evaluations of carbon stocks that consider ecological, climatic, and human factors, deep learning approaches can also combine different information (Xu et al., 2025), such as rainfall patterns, vegetation indices, and land tenure records.

It is impossible to exaggerate the importance of this development. Verifying and valuing carbon storage in ecosystems is essential for compliance markets and voluntary offset programs in a world with carbon constraints. Carbon credits run the risk of becoming unsubstantiated claims in the absence of reliable data. However, nature-based initiatives can offer the legitimacy and openness needed for funding, legislation, and public confidence with AI-powered monitoring. AI turns into the infrastructure that makes it possible to measure the invisible, such as determining how much carbon is sequestered, for how long, and under what circumstances.

AI serves as a link between ecological processes and climate policy in this sense, rather than just being an analytical tool. It enables us to see nature as an active force for planetary healing as well as a passive backdrop for human progress. And in doing so, it assists us in giving the biological systems that subtly support us a scientific, monetary, and ethical value. Measuring and modelling these carbon flows becomes crucial as land-based sequestration becomes a part of global climate efforts. With its ability to make the invisible visible and transform natural carbon sinks into verified climate solutions, AI is quickly becoming essential to this endeavour.

12.2 Below the Surface: AI for Soil Carbon Monitoring

Soil organic carbon (SOC) is an unseen yet powerful climate asset that sits beneath the surface of every terrain. SOC, a vital component of soil fertility and health, is essential to global carbon cycles because it stores CO_2 in stable organic matter, absorbs it through plant photosynthesis, and protects ecosystems from climate extremes (Gerke, 2022). However, soil carbon has long been challenging to quantify, track, and manage on a large scale, despite its significance. Because soil carbon cycles occur in large, diverse terrains, traditional methods—like core sampling and lab analysis—are time-consuming, costly, and impractical. Here, AI is transforming the field by addressing the problem of monitoring carbon stored underground with accuracy, scalability, and predictive capacity (Fig. 12.1).

By identifying trends in a variety of environmental variables, machine learning models are currently being utilized to predict soil organic carbon over wide regions. These models leverage correlations between observed soil carbon values and a variety of input variables, including topography, rainfall patterns, temperature gradients, vegetation indices collected from satellites, and history land use (Yu et al., 2025), rather than solely depending on direct sampling. Researchers and land managers can see where carbon is concentrated, how it varies over time, and where it might be improved through better practices thanks to AI models that, once trained, can accurately forecast soil carbon concentrations across unsampled locations.

The integration of several data sources is essential to this strategy. Narrow portions of the electromagnetic spectrum are captured by hyperspectral satellite imaging, which is especially useful for identifying minute variations in soil composition and vegetation cover. These spectral signals give AI systems rich training data when paired with ground-truth datasets from georeferenced observations and soil cores (El Bouanani et al., 2025). Model dependability is further improved by public soil datasets, such as national inventories, the FAO's Harmonized World Soil Database, and ISRIC's SoilGrids (Batjes, 2022). AI systems get increasingly sophisticated as these datasets expand, becoming capable of adjusting to certain soil types, climates, and management methods (See Fig. 12.2).

Fine-resolution soil carbon mapping is the end outcome, which can direct a variety of land-based climate solutions. These maps assist farmers in identifying carbon-depleted areas that can be used for decreased tillage, compost addition, or cover crops in regenerative agriculture. Soil carbon maps identify regions with high sequestration potential, which helps guide site selection and restoration tactics in afforestation and reforestation initiatives. AI-powered monitoring in rangelands and grazing systems aids in evaluating how native grass recovery or rotational grazing affect soil carbon dynamics, guaranteeing that management strategies are in line with carbon and productivity targets.

These capabilities are already being applied in diverse geographies. Researchers and agricultural organizations are utilizing AI to estimate SOC across cropping zones and degraded lands in Australia, a nation with significant soil and climate variability. These models aid in tracking advancements under soil carbon credit programs and

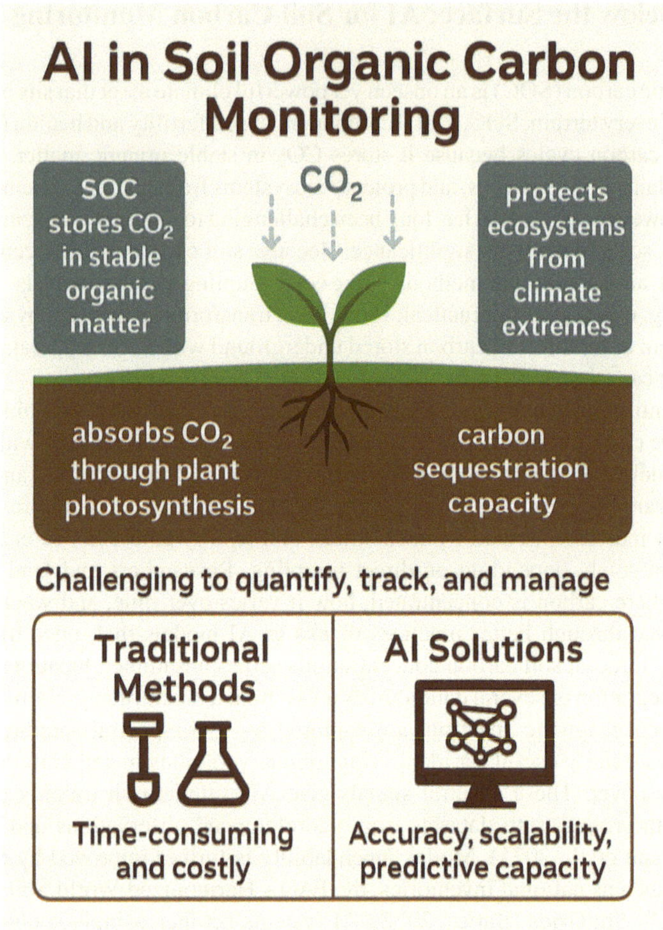

Fig. 12.1 AI revolutionizes soil organic carbon monitoring, replacing slow, costly methods with precise, scalable, predictive tools—enhancing our ability to measure carbon sequestration capacity and protect ecosystems from climate extremes beneath every landscape

prioritizing regions for carbon farming. Machine learning models trained on satellite images and climate data are assisting governments and non-governmental organizations in estimating soil carbon stocks and designing interventions for food security and land restoration in sub-Saharan Africa, where field data is limited but land degradation is pervasive. AI is being used to model SOC under various land uses, from native vegetation to soy cultivation and pasture systems, in Brazil's cerrado biome—a huge tropical savanna under pressure from deforestation and intensive agriculture (Rodrigues et al., 2022)—in order to inform land management and policy (See Fig. 12.3).

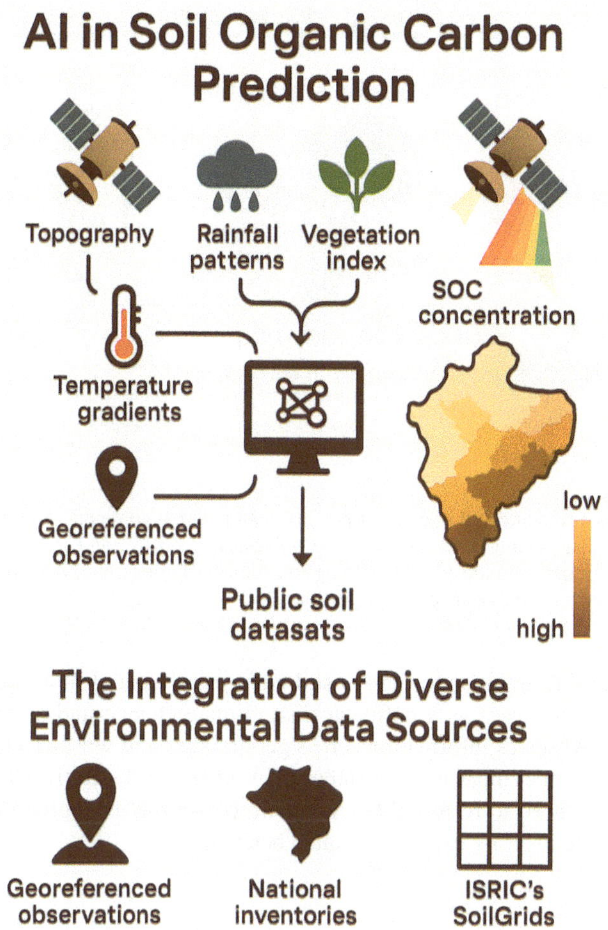

Fig. 12.2 AI-powered models integrate diverse environmental data—topography, rainfall, vegetation indices, temperature, satellite imagery, and public soil datasets—to predict soil organic carbon distribution. This approach enables precise mapping, monitoring, and targeted land management strategies to enhance carbon sequestration and support climate-resilient ecosystems

In this regard, AI's versatility is one of its advantages. It is possible to customize models to fit certain agroecological zones, land uses, and soil types (Mishra, 2025). As management regimes change or new data becomes available, they can be retrained. Additionally, they can be integrated into easily navigable platforms, allowing farmers, local government representatives, and extension agents to access and act upon high-resolution soil carbon insights without requiring sophisticated technical knowledge.

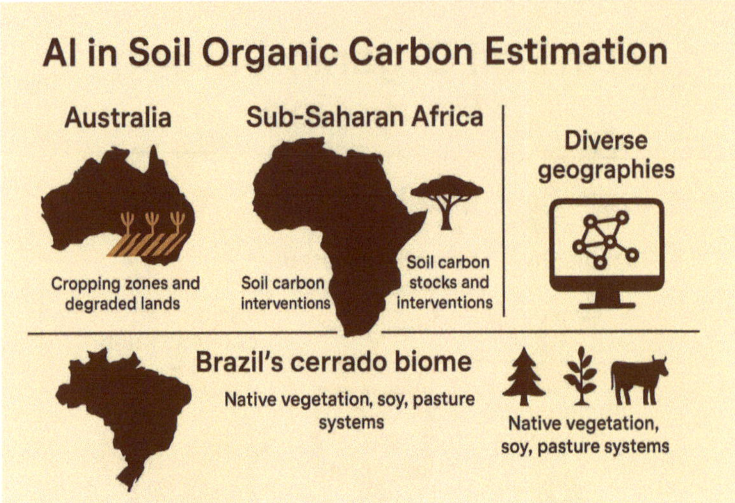

Fig. 12.3 AI is advancing soil organic carbon estimation across continents—mapping cropping zones in Australia, guiding land restoration in Sub-Saharan Africa, and informing sustainable management in Brazil's cerrado biome—helping governments, farmers, and NGOs prioritize carbon farming, food security, and biodiversity protection

The capacity to track and improve soil carbon is essential in a warming world where soil health serves as a buffer against climate change and a foundation for food security. AI scales information across geographies and seasons, enhancing and supplementing direct measurement rather than replacing it. AI gives land stewards the insight they need to restore the land on which we rely for agriculture, climatic stability, and life itself by exposing what is beneath.

12.3 Forests and Biomass: AI for Canopy Intelligence and Carbon Stock Estimation

The biomass of forests—trunks, branches, leaves, and roots—as well as the soils beneath them store enormous amounts of carbon, making them among of the planet's most potent carbon sinks. However, as the global climate crisis worsens, forests are viewed as both increasingly vulnerable frontlines and havens for biodiversity and climate mitigation. Global efforts to fulfill climate targets are threatened by the release of this stored carbon back into the atmosphere due to illegal logging, forest fires, land change, and degradation. We must first comprehend forests—at scale, in detail, and across time—in order to properly protect, restore, and account for them. In order to access this forest wisdom, AI is now essential. This will enable us to measure carbon, identify disturbances, and predict changes with previously unheard-of speed and precision.

The integration of AI with remote sensing is at the heart of this change. Rich, three-dimensional information that can record biomass density, leaf area index, canopy height, and forest structure across large areas are provided by satellite imaging, synthetic aperture radar (SAR), and light detection and ranging (LiDAR) technologies (Kaasalaninen et al., 2015). However, the amount and complexity of this data necessitate analytical skills that go beyond conventional approaches. The ability to extract significant patterns from this data—such as accurately monitoring degradation, mapping changes in forest cover, and predicting above-ground biomass—is made possible by AI, especially deep learning and computer vision algorithms.

For the assessment of forest carbon, LiDAR in particular is revolutionary. It creates high-resolution 3D representations of ground surfaces and forest canopies using laser pulses. By examining tree height, crown breadth, and canopy structure, AI models trained on LiDAR data and field observations can forecast forest biomass. Assessments of the carbon stock over whole forest landscapes are made possible by scaling these estimates with satellite data. AI streamlines this procedure, eliminating the need for expensive and time-consuming ground surveys while preserving or even increasing accuracy.

AI is essential for tracking the dynamics of forests throughout time, even beyond carbon quantification. By comparing satellite imagery over months and years, change detection algorithms can detect regrowth, degradation, afforestation, and deforestation. AI can help discern between seasonal leaf loss and land clearing by differentiating between human and natural changes. Long before they worsen, it can also identify indicators of degradation that are imperceptible to the human eye, including disease outbreaks or canopy thinning from selective logging. AI-enhanced remote sensing aids in the evaluation of burn scars, the classification of fire intensity, and the monitoring of post-fire regeneration in regions that are prone to fire.

Fighting encroachment and illicit logging is one of the most significant applications of AI in forest carbon monitoring. AI may identify anomalous activities, including abrupt canopy loss or the construction of new roads in protected regions, by evaluating satellite feeds almost instantly. This allows authorities to be notified before more damage is done. In areas like the Amazon and Congo Basin, where enforcement capacity is constrained and deforestation can rapidly worsen, this rapid response capability is particularly vital. By tracking smoke plumes, heat signatures, and dryness indices, AI can also aid in early fire identification and expedite the deployment of firefighting resources.

The foundation of climate finance mechanisms such as REDD + (Reducing Emissions from Deforestation and Forest Degradation) are MRV (Measurement, Reporting, and Verification) systems, which are directly fed by these capabilities. AI-enhanced forest monitoring offers the transparency and legitimacy required for nations and project developers taking part in carbon markets to precisely quantify carbon credits. AI lowers transaction costs and increases confidence in emission reduction claims by automating biomass estimation and change detection. Additionally, by enabling local organizations and Indigenous groups to record forest stewardship, open-access AI platforms can validate their role as stewards of carbon-rich ecosystems and qualify them for carbon funding.

The way AI is changing forest carbon accounting is demonstrated by a number of case studies. AI technologies created by groups like MapBiomas and Global Forest Watch employ machine learning and satellite data to track deforestation and forest degradation in the Amazon almost instantly. In order to estimate biomass across large, isolated areas of rainforest that were previously unmapped or incorrectly classified, researchers in the Congo Basin are combining AI with LiDAR and field data. AI is being used in Southeast Asia, especially in Indonesia and Malaysia, to monitor peatland forests, which are among the planet's densest carbon sinks, and identify any illegal burning or drainage that would cause large emissions.

These examples demonstrate that forests are no longer data-blind places. Thanks to AI, we can now keep an eye on their health, see how much carbon they contain, and recognize how they help stabilize the climate. AI is providing us with the means to safeguard forests intelligently, urgently, and strategically because they are living carbon systems rather than just landscapes.

12.4 Toward Transparent Nature-Based Solutions and Climate Accountability

Afforestation, soil carbon improvement, rewilding, and regenerative agriculture are just a few of the NbS for climate mitigation that are gaining traction worldwide, but their legitimacy is also coming under increased scrutiny. Although there is a lot of promise in these land-based methods, their use in national net-zero plans and voluntary carbon markets has sparked concerns about measurement precision, sustainability, and morality. How can we confirm that a tonne of carbon stored in a field or forest is genuine, sustainable, and extra? What measures make sure that local communities gain from carbon projects rather than incur costs? Furthermore, how can investors and governments be sure that nature's carbon sinks are climate assets in reality rather than only empty promises?

AI is becoming a vital ally in bridging these gaps in verification. AI contributes to ensuring that nature-based climate actions are founded on facts rather than conjecture by improving our capacity to measure, track, and predict carbon deposited in ecosystems—both above and below ground. It gives systems that have long struggled with subjectivity and a lack of data openness. AI tools add a new level of legitimacy to carbon accounting, whether it is through anomaly detection that identifies illegal logging, machine learning models that forecast soil organic carbon, or satellite study of forest regrowth.

But data accuracy alone is not enough. The morality of land-based carbon offsets is still complicated and up for debate, especially when it comes to those offered for sale on international voluntary markets. Permanence is one of the primary issues: how long can we anticipate that carbon stored in soils or trees would stay sequestered, particularly in the face of pressures from pests, fire, climate change, and changing land uses? Another is leakage: if systemic measures are not taken, attempts to preserve one forest may merely push deforestation to another area. Most importantly, NbS projects frequently touch on marginalization histories and disputed land rights (Anguelovski & Corbera, 2023). Who decides what happens and who benefits when land is included in a carbon scheme? Better monitoring can be facilitated by AI, but these moral conundrums cannot be resolved unless AI is integrated into inclusive, transparent, and equitable governance frameworks.

The open-source movement becomes crucial at this point. Open access to AI tools, algorithms, and training data promotes responsibility, trust, and cooperative innovation. Governments, non-governmental organizations, local communities, and watchdog groups can access and analyze carbon data via public platforms such as Global Forest Watch, SoilGrids, and OpenET without the need for costly consultants or proprietary software. By enabling users to check assertions, track developments, and share their own observations, these technologies help create a more inclusive carbon stewardship paradigm. By revealing exaggerated carbon claims and guaranteeing that offsets accurately represent ecosystem performance, open-source AI can help stop market manipulation.

Furthermore, establishing climate accountability on a large scale requires AI-powered openness. There is an urgent need for standardized, reliable, and verifiable carbon data as more nations include land-use carbon in their nationally determined contributions (NDCs) under the Paris Agreement (Mor & Ghimire, 2022). More regular and detailed reporting is made possible by AI, which also supports adaptive management and raises the legitimacy of climate pledges. AI is therefore more than just a technological accessory; it is the foundation of the systems that monitor progress, allocate funds, and guide policy in global climate regulation.

The ultimate goal of AI in nature-based solutions is to enhance ecological knowledge and lived experience rather than to replace it. It aids in illuminating the invisible labour of nature, such as the carbon absorbed by roots, the organic matter accumulating in soils, and the regeneration of forest canopies following disturbance (Mandal & Banik, 2025). By turning the incomprehensible into the understandable, it empowers communities, practitioners, and policymakers to take decisive action. Nature has always influenced the environment in a subtle but significant way. We can now start rigorously measuring that contribution, rewarding it with investment, and precisely protecting it thanks to AI. By doing this, we create a future where ecosystems are valued for what they contribute as well as what they protect—a habitable planet—in addition to creating stronger carbon markets.

References

Anguelovski, I., & Corbera, E. (2023). Integrating justice in nature-based solutions to avoid nature-enabled dispossession. *Ambio, 52*(1), 45–53.

Barber, G., Edwards, A., & Zander, K. K. (2024). Nature to the rescue: Past drivers and future potential of the Australian land-based carbon offsets market. *Climate Policy, 24*(5), 601–616.

Batjes, N. H. (2022). Overview of procedures and standards in use at ISRIC WDC-Soils. *Published by ISRIC World Soil Information, Wageningen, Netherlands.*

Cheshmehzangi, A. (2022). Towards sustainable urban green infrastructures. In *Green infrastructure in Chinese cities* (pp. 495–505). Springer Nature Singapore.

Cheshmehzangi, A., Dawodu, A., & Butters, C. (2018, November). Greening urban housing: The impact of green infrastructure on household energy-use reductions for cooling. In *Climate and Energy Protection in the EU and China: 5th Workshop on EU-Asia Relations in Global Politics* (pp. 95–101). Springer International Publishing.

Daba, M. H., & Dejene, S. W. (2018). The role of biodiversity and ecosystem services in carbon sequestration and its implication for climate change mitigation. *Environmental Sciences and Natural Resources, 11*(2), 1–10.

El Bouanani, N., Laamrani, A., Hajji, H., Bourriz, M., Bourzeix, F., Ait Abdelali, H., El-Battay, A., Amazirh, A., & Chehbouni, A. (2025). Estimating soil attributes for yield gap reduction in Africa using hyperspectral remote sensing data with artificial intelligence methods: An extensive review and synthesis. *Remote Sensing, 17*(9), 1597.

Fernández-Romero, M. L., Lozano-García, B., & Parras-Alcántara, L. (2014). Topography and land use change effects on the soil organic carbon stock of forest soils in Mediterranean natural areas. *Agriculture, Ecosystems & Environment, 195*, 1–9.

Gerke, J. (2022). The central role of soil organic matter in soil fertility and carbon storage. *Soil Systems, 6*(2), 33.

Kaasalainen, S., Holopainen, M., Karjalainen, M., Vastaranta, M., Kankare, V., Karila, K., & Osmanoglu, B. (2015). Combining lidar and synthetic aperture radar data to estimate forest biomass: Status and prospects. *Forests, 6*(1), 252–270.

Li, L., Chan, F., & Cheshmehzangi, A. (2023). Nature-based solutions and sponge city for urban water management. *Adapting the built environment for climate change* (pp. 371–402). Woodhead Publishing.

Loria, N., Lal, R., & Chandra, R. (2024). Handheld in situ methods for soil organic carbon assessment. *Sustainability, 16*(13), 5592.

Mandal, S., & Banik, G. C. (2025). Forest degradation and its impact on soil carbon. In *Forest degradation and management: An Indian perspective* (pp. 207–225). Springer Nature Switzerland.

Mishra, U. (2025). Harnessing AI technologies for sustainable agricultural practices: Innovations in soil analysis and crop management. *Biology, Engineering, Medicine and Science Reports, 11*(1), 9–13.

Mor, S., & Ghimire, M. (2022). Transparency and nationally determined contributions: A review of the Paris Agreement. *Interdisciplinary Journal of Economics and Business Law, 11*(2), 106–119.

Rodrigues, A. A., Macedo, M. N., Silvério, D. V., Maracahipes, L., Coe, M. T., Brando, P. M., Shimbo, J. Z., Rajão, R., Soares-Filho, B., & Bustamante, M. M. (2022). Cerrado deforestation threatens regional climate and water availability for agriculture and ecosystems. *Global Change Biology, 28*(22), 6807–6822.

Seddon, N., Chausson, A., Berry, P., Girardin, C. A., Smith, A., & Turner, B. (2020). Understanding the value and limits of nature-based solutions to climate change and other global challenges. *Philosophical Transactions of the Royal Society B, 375*(1794), 20190120.

Twyman, C., Smith, T. A., & Arnall, A. (2015). What is carbon? Conceptualising carbon and capabilities in the context of community sequestration projects in the global South. *Wiley Interdisciplinary Reviews: Climate Change, 6*(6), 627–641.

Xu, W., Cheng, Y., Luo, M., Mai, X., Wang, W., Zhang, W., & Wang, Y. (2025). Progress and limitations in forest carbon stock estimation using remote sensing technologies: A comprehensive review. *Forests, 16*(3), 449.

Yu, Z., Deng, X., Cheshmehzangi, A., & Gao, Y. (2025). Assessing future land use demands in response to land degradation risk and Socio-Economic impacts. *Ecological Indicators, 175*, Article 113529.

Chapter 13
Eyes on the Forest: AI for Real-Time Deforestation Detection and Land Protection

13.1 Forests Under Siege: The Scope and Urgency of Deforestation

The rate at which forests are disappearing is concerning. Because of a combination of natural disturbances and human activity, the world loses more than 10 million hectares of tree cover annually (Dar et al., 2022), which is about the size of Portugal. The main driver is still agriculture, especially in tropical areas where forests are being cut down for oil palm plantations, soy production, and cow grazing. Large areas of forest are ripped out by mining activities in search of rare earths and precious metals. Both legal and illicit logging contribute to the world's demand for paper and lumber. Roads, pipelines, and urbanization are examples of expanding infrastructure that slashes into ecosystems that were once untouched. Furthermore, deforestation is both a cause and an effect of climate-induced wildfires, which are becoming more frequent and intense, particularly in sensitive regions like the Amazon, Siberia, and Borneo.

This loss has effects that go well beyond trees. Approximately one-third of the annual global CO_2 emissions from fossil fuels are absorbed by forests, making them essential carbon sinks (Chaudhry, 2019). Carbon sinks become carbon sources when forests are cut down, burned, or otherwise damaged because they release stored carbon back into the atmosphere (Cheshmehzangi & Tang, 2022; Yu et al., 2023). This contributes to drought, soil erosion, and agricultural decline in addition to hastening climate change and upsetting local weather patterns. Since more than half of the terrestrial species on Earth are found in tropical forests, many of which are unique to this planet, deforestation is also a problem for biodiversity. It's also a matter of justice. As their lands are encroached upon, indigenous tribes and peoples who depend on forests—often the best custodians of the land—are uprooted, ostracized, or criminalized.

© The Author(s), under exclusive license to Springer Nature Singapore Pte Ltd. 2025

A. Cheshmehzangi, *AI Agency in Climate Emergency*, Urban Sustainability,
https://doi.org/10.1007/978-981-95-4192-8_13

Deforestation is still mostly unabated despite decades of international accords, commitments, and conservation efforts (Yu et al., 2022). The limitations of conventional forest monitoring technologies have been a significant obstacle to a successful response. Since many forest loss alarms are based on antiquated satellite systems with poor spatial resolution, it can be challenging to identify early indicators of encroachment or small-scale illicit logging (Pandey et al., 2022). Despite being crucial, field-based reporting is frequently delayed, expensive, and risky, particularly in isolated or conflict-prone areas. Damage is frequently done by the time deforestation is documented and confirmed. A crucial gap in the global climate and conservation efforts is the delay between forest loss and action (See Fig. 13.1).

A real-time, high-resolution, publicly available system for monitoring forest change is desperately needed. This system should be able to spot early warning indicators, identify disturbances as they occur, and enable communities, governments, and civil society to act swiftly and efficiently. AI presents a revolutionary advancement in this regard.

By significantly increasing the speed, scope, and precision of land change identification, AI is changing the potential of forest monitoring (Ali et al., 2025). AI models can process terabytes of imagery in almost real-time by utilizing large datasets from commercial platforms like Planet and high-resolution satellite constellations like Sentinel and Landsat (Akhtar & Rawol, 2025). Patterns of change, such as clear-cuts, road construction, canopy thinning, and burn scars, can be identified by machine learning algorithms trained on annotated forest footage. AI systems can operate continuously, identifying possible deforestation events within days or even hours of occurrence, in contrast to traditional systems that rely on human interpretation on a periodic basis.

AI is unique not only because of its speed but also because of its sensitivity. Deep learning algorithms can differentiate between temporary logging roads and permanent land conversion, as well as between seasonal leaf drop and actual deforestation (Wang et al., 2025). They are able to identify subtle types of degradation, including selective logging, understory clearing, or forest edge fragmentation, that might not be visible as loss in lower-resolution systems. These subtleties are important because forest degradation is a major but underreported contribution to emissions and biodiversity loss, and it frequently occurs before complete deforestation.

The accessibility that AI makes possible is equally significant. A wider spectrum of people, including Indigenous land defenders, journalists, park rangers, environmental NGOs, and local governments, are now able to access AI-powered forest monitoring tools thanks to cloud computing and open-source platforms. By democratizing access to forest data, programs like Rainforest Connection, MapBiomas (Taylor et al., 2020), and Global Forest Watch are transforming algorithms and satellites (Bahar & Wicaksono, 2021) into tools for local empowerment and responsibility (See Fig. 13.2).

The methods for preventing deforestation must change as it becomes more complex and widespread. Seeing land change as it occurs is now a planetary imperative rather than a technical luxury. AI gives us the vision, speed, and scale we need to save what's left of the world's forests, and to do it quickly enough to have an impact.

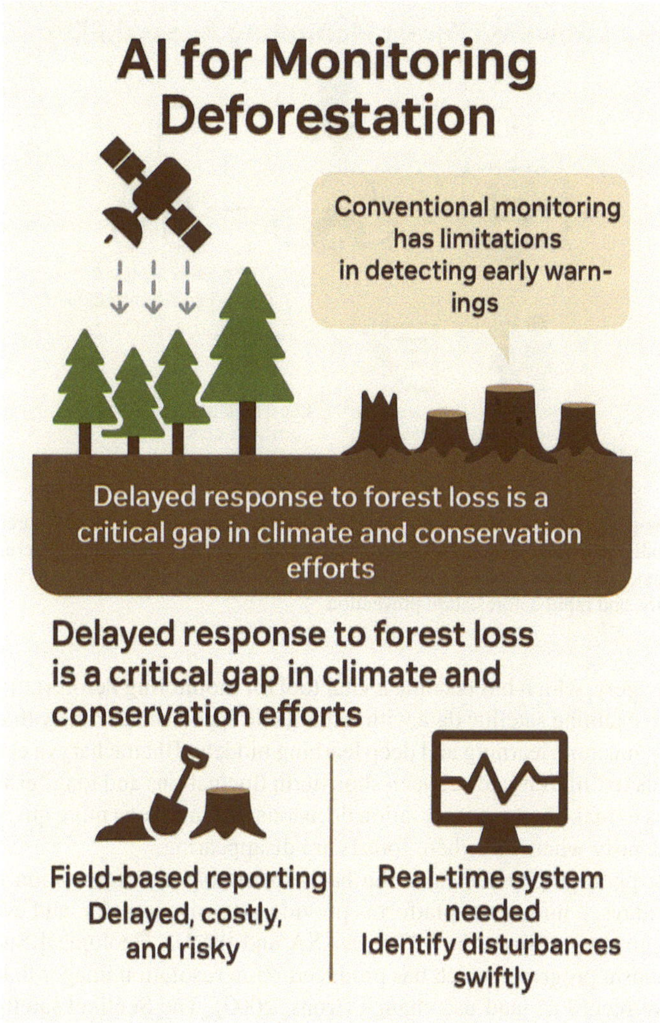

Fig. 13.1 AI-driven forest monitoring bridges the critical gap between deforestation detection and action, overcoming delays from low-resolution satellites and costly field reporting. Real-time, high-resolution systems enable swift identification of disturbances, empowering timely climate and conservation responses worldwide

13.2 Seeing from Space: How AI Analyzes Satellite Imagery for Deforestation

Satellites, which are hundreds of kilometers above the ground, take daily pictures of the shifting terrain of our planet, which is made up of layered mosaics of water, forest cover, and human activities. The true innovation is in how we analyze this

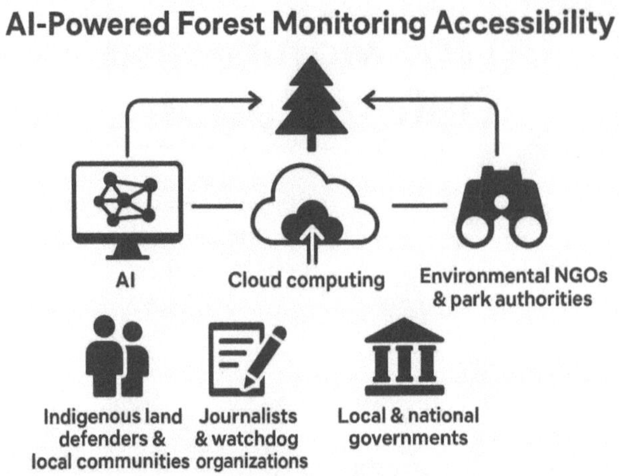

Fig. 13.2 AI-powered forest monitoring, enabled by cloud computing and open-source platforms, empowers Indigenous land defenders, journalists, park rangers, NGOs, and local governments to act quickly. Democratized access transforms satellites and algorithms into powerful tools for protection, accountability, and rapid deforestation prevention

flood of imagery, which has become a vital tool for monitoring deforestation. We are now able to examine satellite data with remarkable speed and accuracy thanks to AI, specifically machine learning and deep learning models (Bhattacharyya et al., 2023). AI allows us to differentiate between short-term fluctuations and long-term changes, allowing us to make better conservation decisions and intervene more quickly. It also helps us identify when and where forests are disappearing.

Satellite photography serves as the basis for AI-based deforestation identification. Nowadays, a number of platforms provide extensive, regular, and ever higher-resolution coverage. Since the 1970s, NASA and the US Geological Survey have run the Landsat program, which has produced 30-m resolution images that provides a long-term record of land use change (Irons, 2000). The Sentinel satellites of the European Space Agency provide free imagery with a resolution of 10–20 m and a few-day revisit frequency. Every day, commercial companies such as Planet Labs set up constellations of CubeSats to take 3- to 5-m-resolution pictures of the Earth's surface (Azami et al., 2022). Continuous forest monitoring at sizes ranging from individual tree plantations to entire biomes is made possible by this rich tapestry of temporal and geographical data (See Fig. 13.3).

Satellites gather the data, but AI interprets it. By examining the spectral and spatial properties of every pixel in a satellite image, machine learning algorithms process and categorize land cover. Random Forests, which categorize land cover types using decision-tree ensembles, and convolutional neural networks (CNNs), which are particularly good at interpreting image-based data, are two of the most popular models. Deep learning techniques may identify intricate elements like the

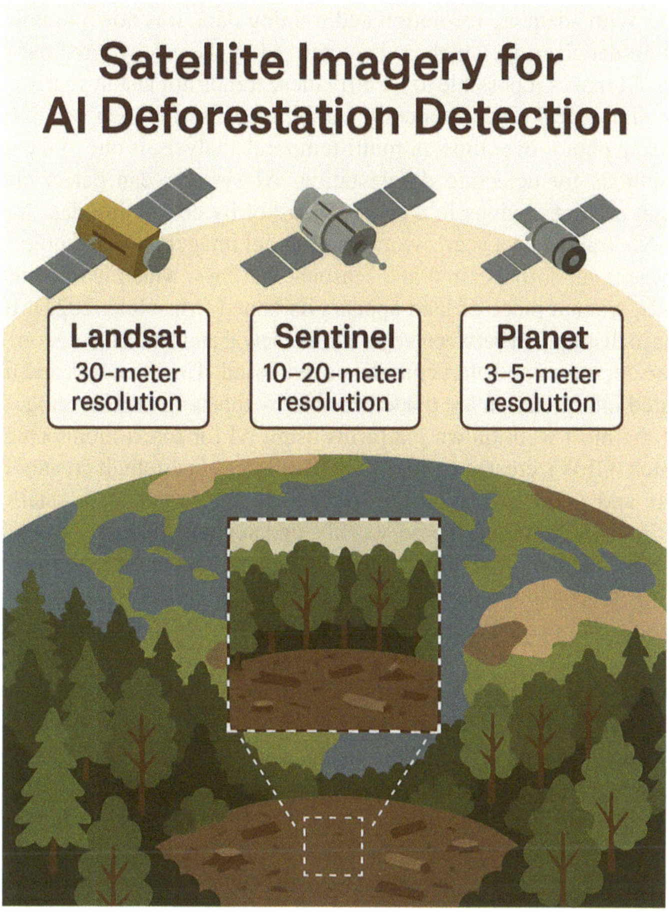

Fig. 13.3 AI-driven deforestation detection relies on satellite imagery from Landsat, Sentinel, and Planet. Their varying resolutions and revisit frequencies enable continuous forest monitoring, from global biomes to individual plantations, empowering timely interventions and informed conservation strategies worldwide

texture of a canopy, the geometry of a road, or the shadow of a fire scar. These techniques are frequently trained on enormous datasets of annotated deforestation events.

A large variety of forest disturbances can be detected by these models. Clear-cutting and selective logging can also cause canopy loss, which is frequently identified by an abrupt shift in the near-infrared reflectance that plant normally emits. The straight geometry and colour contrast of newly constructed paths through forested areas are indicators of road encroachment, which is a prelude to deforestation. AI algorithms are able to recognize the unique thermal and spectral fingerprints of burn scars from multispectral data, which are created by forest fires or slash-and-burn agriculture. With adequate resolution and training data, it is now possible to follow even smallholder clearing, which is characterized by the patchy, irregular removal of vegetation. AI makes it possible to identify these trends not just in yearly evaluations but also in almost real time, frequently within a few days of their occurrence.

Comparing photos over time, or multi-temporal analysis, is one of the most effective AI methods for detecting deforestation. AI systems can detect changes and patterns that are not obvious in a single snapshot by combining data from several dates. A CNN trained on a year's worth of Sentinel imagery, for example, can distinguish between real deforestation and seasonal leaf loss, which is frequent in deciduous woods, when a piece of land appears darkened (Masolele, 2023). By training algorithms to distinguish between vegetation renewal and land conversion, false positives can be decreased and alert confidence increased. Timely action and much more sophisticated interpretation are made possible by this temporal layering.

One of the most well-known platforms using AI for forest monitoring is Global Forest Watch (GFW), created by the World Resources Institute. It creates Tree Cover Loss alerts and other indications by fusing AI algorithms with satellite images from Sentinel and Landsat. These weekly updates and publicly accessible alerts give governments, non-governmental organizations, media outlets, and Indigenous communities access to timely and useful information. GFW distinguishes between anthropogenic and natural causes of loss by combining CNNs and Random Forest models.

Another notable example in the Amazon is the Monitoring of the Andean Amazon Project (MAAP) (Finer et al., 2024a, 2024b). It tracks deforestation hotspots in Ecuador, Brazil, Colombia, and Peru using machine learning and high-resolution Planet imagery (Finer & Mamani, 2020). Finding illegal gold mining areas, monitoring the expansion of agriculture, and evaluating the effects of infrastructural development have all benefited greatly from MAAP's findings. Real-time data inputs from the platform enable local law enforcement to take prompt action, frequently stopping additional forest encroachment.

Synthetic Aperture Radar (SAR) is becoming an increasingly important data source in Southeast Asia, particularly in tropical, overcast areas where optical vision is frequently obstructed. SAR sensors can reliably identify tree loss, plantation expansion, and peatland disturbance because they can pierce cloud cover and detect changes in ground structure and moisture. AI models built on SAR data, like those in Malaysia and Indonesia, are assisting in the monitoring of rainforest logging and warning authorities of illicit clearing in protected peat zones (See Fig. 13.4).

Forest governance is changing because of these technologies working together. Deforestation is no longer a covert operation because to AI's ability to analyze satellite imagery quickly and extensively. Almost as it occurs, forest destruction becomes apparent, allowing for prompt intervention and accountability of those responsible. This potential is more than just a technological advancement in a time when every hectare of forest matters; it is a lifeline for the ecosystems throughout the planet.

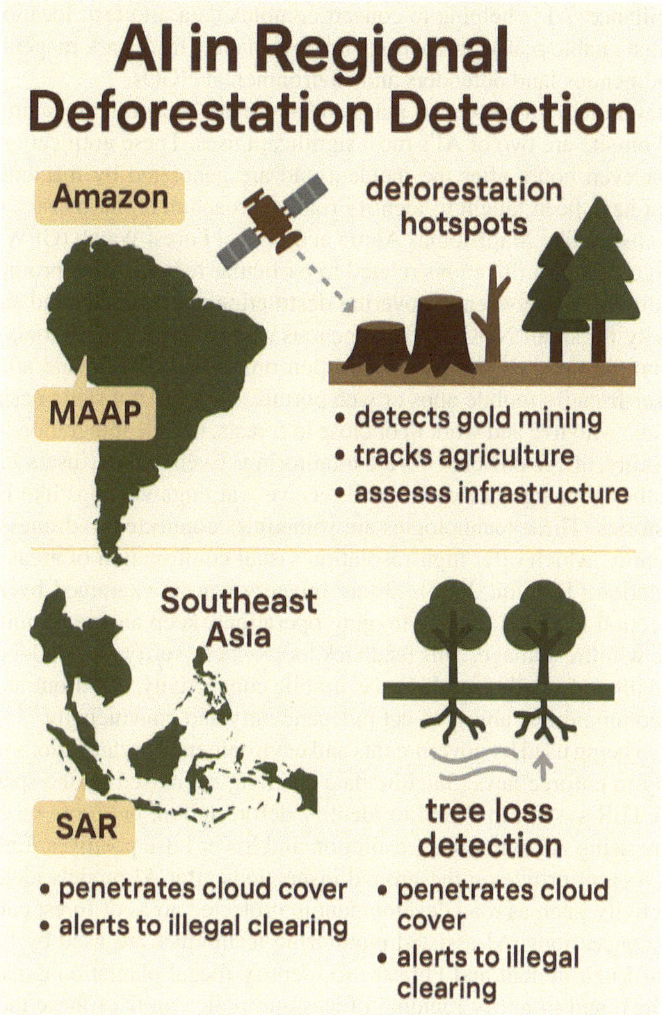

Fig. 13.4 MAAP in the Amazon and SAR in Southeast Asia use AI and advanced imagery to detect deforestation, monitor agriculture, and alert authorities to illegal activities, even through persistent tropical cloud cover

13.3 Actionable Intelligence: Enforcement, Community Protection, and Governance

Timely information is becoming essential rather than optional as deforestation grows more dynamic, decentralized, and frequently illegal. For individuals on the front lines of forest conservation, AI is adding a new layer of actionable intelligence when combined with digital tools and satellite photos. This is about empowerment, not only surveillance. AI is helping to convert complex data into fast, location-specific warnings that enable real-world action and governance, from park rangers in distant areas to Indigenous land defenders and environmental NGOs.

The creation of high-resolution maps and automatic deforestation alerts that may be utilized on-site are two of AI's most significant uses. These notifications are sent out days or even hours after the incident and are generated by machine learning models that have been taught to identify road encroachment, burn scars, or canopy loss. On websites like MapBiomas Alerta and Global Forest Watch (GFW), individuals can sign up for notifications related to particular regions. This promptness can mean the difference between uncovering destruction after the fact and stopping an illicit activity for small NGOs and Indigenous communities. These tools are living systems that are updated and improved upon on a regular basis and are available through user-friendly mobile apps or web portals. They are not static dashboards.

For people who live and work in or close to forests, mobile integration is essential to the viability of AI-powered forest monitoring. Even offline, users can upload reports, gather geotagged evidence, and receive warnings via apps like ForestLink and Terrastories. These technologies are sometimes connected to drones owned by the community, which offer high-resolution visual confirmation of areas suspected of deforestation (Termini, 2025). Drone imagery can be examined by AI-trained object detection models to verify logging operations, keep an eye on mining sites, or evaluate wildfire damage. This feedback loop—alert, verify, respond—created by the combination of satellite intelligence, mobile connectivity, and drone surveillance enhances communities' ability to act independently and convincingly.

AI is also being used by governments and environmental organizations to improve their ability to enforce laws. Satellite data has long been used by the space agency INPE's DETER system in Brazil to identify deforestation, but more recent collaborations are using AI to improve resolution and lower false positives. Enforcement teams can then prioritize on-the-ground inspections after AI models identify questionable activity, such as road development in protected areas or forest cutting close to mining concessions. AI-assisted monitoring techniques are used by Indonesia's Ministry of Environment and Forestry to identify illegal plantation expansion and peatland fires, and to notify regional offices and police units of these incidents. In addition to increasing accuracy, AI makes triage possible, guaranteeing that limited law enforcement resources are allocated to the most pressing concerns.

Another striking example is Central Africa, where large areas of rainforest in the Congo Basin are being monitored using AI-based platforms. Due to a lack of infrastructure and enforcement capabilities, governments and non-governmental organizations are using tools such as Rainforest Foundation UK's MappingForRights and WRI's Forest Atlas to monitor concessions, verify protected area borders, and assist with anti-logging efforts. The political stakes are high here since many regions are both contested Indigenous land claims and sites of ecological significance. AI promotes transparency by offering time-stamped, verifiable proof of compliance or incursion that may be utilized in court or for international reporting.

AI is becoming increasingly involved in governance and participatory conservation in addition to enforcement. By recording land use, identifying encroachment, and highlighting the stewardship methods that have protected forests for generations, it offers resources that promote Indigenous land rights. Local communities and Indigenous peoples can more successfully defend their lands, assert their rights, and take part in decision-making when they have access to the same data as governments or businesses. With the help of AI to support their expertise with empirical data, community-based forest monitors are becoming more and more acknowledged as co-managers of conservation areas in places like the Amazon and portions of Southeast Asia.

Lastly, supply networks are becoming more accountable thanks in large part to AI. AI enables businesses and watchdog organizations to track down the origin of deforestation-linked products by connecting deforestation warnings to the production of agricultural commodities like soy, cattle, or palm oil. AI is used by platforms such as Trase and Satelligence to monitor changes in land usage close to farms or plantations and evaluate adherence to zero-deforestation pledges. In addition to keeping bad actors out of supply chains, this intelligence is also being utilized to incentivize people that preserve forest cover or embrace sustainable practices.

The potential of AI to disseminate insight, transform data into action, and provide those closest to the forest with capabilities previously only accessible by distant analysts is what gives it its real power in deforestation governance. AI is establishing a new geography of vigilance where forest protection is quicker, more informed, and more inclusive, whether in the areas of enforcement, community protection, or corporate accountability.

13.4 The Politics of Monitoring: Data Ethics, Accountability, and Global Solidarity

A new level of complexity—not technological, but political and ethical—emerges as AI-driven deforestation monitoring becomes more integrated into global conservation and climate initiatives. Significant responsibility accompanies unparalleled visibility. The problems of who controls the data, who determines its interpretation, and whose interests it eventually serves are brought up by the capability of real-time land change observation from space. AI has the potential to democratize access to forest intelligence, but unless conscious efforts are made to ensure transparency, permission, and fair involvement, it can also concentrate power in the hands of a small number of players.

When AI is used in satellite surveillance without established ethical guidelines, there are concerns involved. The same instruments that are used to track illicit deforestation can also be used to criminalize Indigenous customs, subsistence farming, and opposition to extractive industries in areas where land rights are contested. Data on deforestation may be abused by governments to support evictions or to greenwash destructive development initiatives. Businesses could select specific information to demonstrate compliance while neglecting more extensive environmental damage.

Without precautions, forest monitoring turns into a tool for control rather than conservation. Unauthorized surveillance can perpetuate the very inequalities that environmental movements aim to correct, particularly in regions with colonial or authoritarian pasts.

Any AI-enabled monitoring system must be built on a basis of openness and shared governance to mitigate these dangers (Cheshmehzangi, 2025). Open access to methodology, datasets, and model assumptions are all part of this. The right to access data, comprehend its usage, and contest its interpretation should be granted to communities whose lands are being watched. In order for affected stakeholders to have a role in what is monitored, how alerts are created, and what happens next, consent needs to be more than just a formality. Reducing false positives in algorithms is only one aspect of ethical forest monitoring; another is putting justice at the heart of the environmental intelligence infrastructure.

In this case, open-source platforms are essential. Platforms like Global Forest Watch, MapBiomas, and Earthrise Alliance lessen reliance on closed ecosystems and proprietary technologies by making satellite analysis tools, machine learning models, and alert systems publicly available. Academic institutions, Indigenous organizations, and civil society groups can all review, refine, localize, and adapt open-source models. They facilitate a more decentralized, transparent monitoring environment and encourage cooperation rather than rivalry. To ensure that local actors have the knowledge, resources, and tools necessary to interact with these systems in a meaningful way, open access must be paired with capacity-building.

If public–private collaborations are designed with equity in mind, they can also lead to responsible innovation. Governments, satellite providers, and IT firms can collaborate to increase the frequency, resolution, and scope of forest monitoring (Fuller, 2006). However, extractive dynamics—where local data is collected without benefit-sharing—must be avoided in collaborations. Forest data systems need to acknowledge and value the contributions of local stewards, much as carbon markets are coming to understand the value of equitable remuneration for ecosystem services. Shared values such as accountability, transparency, and climate justice must serve as the foundation of any partnership.

Lastly, AI's potential for protecting forests rests in unity rather than surveillance. It equips us to recognize deforestation as it occurs, but the real strength is in putting that awareness into action. Forest alerts are only effective if they force businesses to alter their operations, enable communities to protect their lands, or spur rapid enforcement. Data isn't enough to stop a bulldozer. It is prevented by group action supported by facts, the law, and moral clarity.

In a time of planetary crisis, AI offers a rare convergence of urgency and capacity. We can now observe environmental devastation in real time and decide to act before the harm is done, rather than after it has already occurred. This intelligence becomes more than just a detecting tool if we use it responsibly, based on morality, inclusion, and international solidarity. In defending the world's woods, it turns into a force for justice, regrowth, and optimism.

References

Akhtar, Z. B., & Rawol, A. T. (2025). Harnessing artificial intelligence (AI) towards the landscape of big earth data: Methods, challenges, opportunities, future directions. *Journal of Geography and Cartography, 8*(1), 10224.

Ali, G., Mijwil, M. M., Adamopoulos, I., & Ayad, J. (2025). Leveraging the internet of things, remote sensing, and artificial intelligence for sustainable forest management. *Babylonian Journal of Internet of Things, 2025*, 1–65.

Azami, M. H. B., Orger, N. C., Schulz, V. H., Oshiro, T., & Cho, M. (2022). Earth observation mission of a 6U CubeSat with a 5-meter resolution for wildfire image classification using convolution neural network approach. *Remote Sensing, 14*(8), 1874.

Bahar, N. H., & Wicaksono, S. A. (2021). How to effectively engage youth in satellite-based tropical forest monitoring?. *Innovative forestry for a sustainable future*, 5.

Bhattacharyya, A., Nambiar, S. M., Ojha, R., Gyaneshwar, A., Chadha, U., & Srinivasan, K. (2023). Machine Learning and Deep Learning powered satellite communications: Enabling technologies, applications, open challenges, and future research directions. *International Journal of Satellite Communications and Networking, 41*(6), 539–588.

Chaudhry, A. A. (2019, June). The role of forests in carbon sequestration to counter air pollution. In *On behalf of Pakistan Engineering Congress.*

Cheshmehzangi, A. (2025). *Urban Innovation and Sustainability: ICT-based interventions in Hainan, China.* Springer Nature.

Cheshmehzangi, A., & Tang, T. (2022). *China's city cluster development in the race to carbon neutrality*. Springer.

Dar, S. A., Nabi, M., Dar, S. A., & Ahmad, W. S. (2022). Influence of anthropogenic activities on the diversity of forest ecosystems. *Towards sustainable natural resources: Monitoring and managing ecosystem biodiversity* (pp. 33–49). Springer International Publishing.

Finer, M., & Mamani, N. (2020). MAAP Synthesis: 2019 Amazon Deforestation Trends and Hotspots. *MAAP Synthesis, 4.* Available from: https://www.amazonconservation.org/tag/synthesis/

Finer, M., Mamani, N., Anderson, C., & Rosenthal, A. (2024a). *Carbon across the Amazon (part 2): Peak Carbon Areas.* Available from https://www.maapprogram.org/category/maps/

Finer, M., Mamani, N., Anderson, C., & Rosenthal, A. (2024b). *Carbon across the Amazon (part 3): Key Cases of Carbon Loss & Gain.* Available from https://www.maaproject.org/author/dcadmin/

Fuller, D. O. (2006). Tropical forest monitoring and remote sensing: A new era of transparency in forest governance? *Singapore Journal of Tropical Geography, 27*(1), 15–29.

Irons, J. R. (2000). Present and future of the Landsat program. *Algorithms for Multispectral, Hyperspectral, and Ultraspectral Imagery VI, 4049*, 122–133.

Masolele, R. N. (2023). *Assessing land use following tropical deforestation: Combining remote sensing and deep learning* (Doctoral dissertation). Wageningen University and Research.

Pandey, S., Kumari, N., Dash, S. K., & Al Nawajish, S. (2022). Challenges and monitoring methods of forest management through geospatial application: A review. *Advances in Remote Sensing for Forest Monitoring*, 289–328.

Taylor, R., Davis, C., Brandt, J., Parker, M., Stäuble, T., & Said, Z. (2020). The rise of big data and supporting technologies in keeping watch on the world's forests. *International Forestry Review, 22*(1), 129–141.

Termini, S. L. (2025). *Landscape of memories: Preserving oral history using volunteered geographic information (VGI) in a web GIS application* (Doctoral dissertation). California State University, Northridge.

Wang, T., Zuo, Y., Manda, T., Hwarari, D., & Yang, L. (2025). Harnessing artificial intelligence, machine learning and deep learning for sustainable forestry management and conservation: Transformative potential and future perspectives. *Plants, 14*(7), 998.

Yu, Z., Deng, X., & Cheshmehzangi, A. (2022). The grain for green program enhanced synergies between ecosystem regulating services in Loess Plateau, China. *Remote Sensing, 14*(23), 5940.
Yu, Z., Deng, X., Cheshmehzangi, A., & Mangi, E. (2023). Structural succession of land resources under the influence of different policies: A case study for Shanxi Province, China. *Land Use Policy, 132*, Article 106810.

Part V
AI for Transparent Carbon Accounting and Climate Finance

This section explores how AI is improving carbon accounting and climate finance's accountability, transparency, and efficiency. By examining supply chains, energy consumption, and satellite data, AI-powered solutions can monitor emissions across industries and provide precise, real-time estimates of carbon footprints. AI is used in the finance sector to forecast exposure to climate change, assess climate investment opportunities, and detect greenwashing concerns. Additionally, by confirming offsets, predicting trends, and identifying fraud, AI improves carbon markets and strengthens the efficacy and credibility of emissions trading. When combined, these applications support believable routes to global net-zero goals and climate action.

Chapter 14
Algorithms of Accountability: AI for Emissions Tracking in a Carbon-Constrained World

14.1 The Carbon Blindspot: Why Emissions Are Hard to Measure

Accurately and reliably quantifying carbon emissions is still one of the biggest challenges in the fight to slow down climate change. The world continues to function under a massive carbon blindspot despite decades of international agreements, commitments, and protocols. Global emissions are the result of a complex network of interconnected, frequently opaque, and rapidly evolving activities, ranging from extensive power grids and industrial complexes to scattered agricultural areas, transportation networks, and supply chains. It has been difficult to comprehend this carbon footprint in real time, let alone at the level of detail necessary for successful intervention. AI is filling this gap by providing previously unheard-of capabilities to monitor, decipher, and influence the unseen gases influencing the future of our planet.

Emissions occur in many different sectors and regions, each with unique difficulties. The majority of the world's CO_2 emissions are caused by burning fossil fuels, making the energy sector the biggest emitter. Steelmaking, chemical refining, and cement manufacture are all intricate processes that include industrial emissions and are frequently obscured by private data. Automobiles, ships, aircraft, and freight networks that cross international borders all produce pollution. Through the use of fertilizer, cattle, and rice farming, agriculture adds methane and nitrous oxide, while changes in land use, especially deforestation and peatland drainage, release enormous amounts of carbon that are hard to detect and much more difficult to precisely measure. The entirety of these emissions is currently not fully and verifiably captured by any one technique or organization.

Despite their importance, traditional carbon accounting systems have significant drawbacks. For international reporting, such as that mandated by the UNFCCC, the majority rely on national greenhouse gas inventories that are periodically compiled (Yona et al., 2022). These inventories are based on estimations, emissions factors, and statistical models, many of which are based on out-of-date or insufficient data. Many

© The Author(s), under exclusive license to Springer Nature Singapore Pte Ltd. 2025

A. Cheshmehzangi, *AI Agency in Climate Emergency*, Urban Sustainability,
https://doi.org/10.1007/978-981-95-4192-8_14

reports are released one to three years after the fact, which makes them unsuitable for providing information for immediate action. Additionally, a large portion of the system is self-reported by businesses and governments, which raises concerns about methodological inconsistencies, openness, and the possibility of selective or underreported disclosure. Another obstacle is fragmentation: supply chain emissions (Scope 3) are frequently left out completely (Borchardt et al., 2025; Wieland & Creutzig, 2025), whereas industrial emissions may be reported independently from land-use emissions.

Emissions opacity in international supply chains is a particularly challenging problem. Products used in one nation frequently produce emissions in another due to the extraction, production, and transportation of raw materials (Küfeoğlu, 2024). However, national inventories usually fail to include these emissions, resulting in a blind spot in climate stewardship. Similarly, although though it makes a substantial contribution to local emissions and land-use change, the informal economy—which includes anything from smallholder agriculture to unregulated mining—operates outside of official reporting systems. The problem is made worse by carbon leakage, which obscures the actual emissions footprint of many economies by moving high-emitting activities over international boundaries to evade regulation.

In light of this, there has never been a more pressing need for scalable, transparent, and real-time carbon accounting. Carbon pricing, regulatory compliance, climate finance, corporate ESG disclosures, and global net-zero commitments all depend on reliable emissions data. Without it, we run the danger of investing in illusory solutions, creating policies based on hunches, or failing to identify important emissions hotspots. The ability to see emissions clearly decides who is held accountable, who receives climate finance, and which solutions are prioritized, making this a governance and justice issue rather than only a technical one.

AI presents a fresh approach. AI can identify trends, close gaps, and produce dynamic emissions estimates at various scales by examining vast and varied datasets, including satellite imagery, IoT sensors, industrial activity logs, trade flows, agricultural models, and more. Thermal anomalies at power plants, shipping routes in marine traffic, fertilizer purchases associated with nitrous oxide release (Ma et al., 2017), or changes in land cover associated with forest carbon loss are examples of proxy indicators from which machine learning algorithms can predict emissions. AI systems can track emissions almost instantly, adapting and updating as circumstances change (See Fig. 14.1).

Most importantly, AI can reveal the invisible. It can reveal emissions scattered throughout the unorganized sector, concealed in subcontracted processes, or incorporated in goods. It can create coherent, interoperable emissions maps by combining disparate data sources. Additionally, it can give communities, businesses, and legislators the knowledge they need to take action now, rather than years from now. Ignorance is no longer an option in a world with carbon constraints (Brandon, 2012). We must first have a clear understanding of what we must reduce to control it. By converting blind spots into baselines, AI is starting to provide that clarity and give us a chance to match emissions monitoring to the scope and urgency of the climate catastrophe (Cheshmehzangi et al., 2025). This entails a change from reactive to

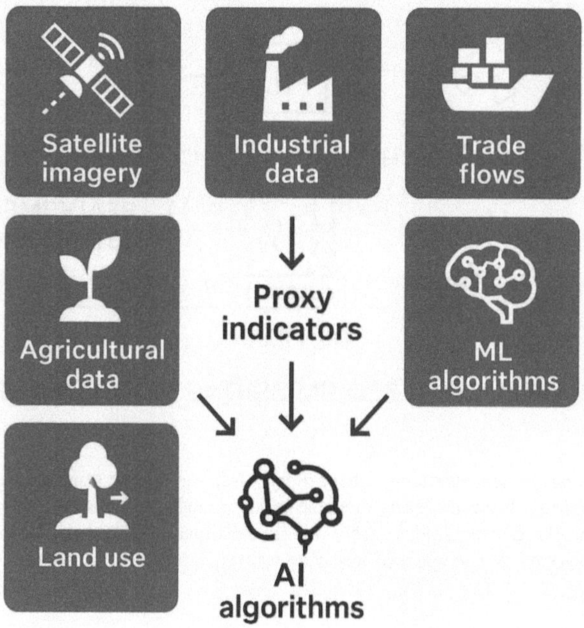

Fig. 14.1 AI-driven emissions monitoring integrates diverse data sources—satellite imagery, industrial records, trade flows, agricultural data, and land use changes—into proxy indicators analyzed by machine learning. This approach reveals hidden emissions, creates coherent maps, and empowers timely climate action by closing data gaps and enabling near-real-time environmental oversight

proactive tactics that foresee and stop high-emission behaviours before they become more serious. For openness to result in real change, these findings must also be included into industry standards, urban planning, and policies.

14.2 Sensing Emissions: Satellite Imagery, Remote Sensing, and AI Integration

Satellites are subtly changing our understanding of carbon emissions on Earth from the high atmosphere. The combination of remote sensing and AI has made previously concealed phenomena—such as methane leaking from landfills, flaring gas from oil fields, and thermal leakage from power plants—more visible. Long constrained by coarse resolution and occasional observations, satellite-based emissions monitoring is about to enter a new age. AI is making it possible to estimate greenhouse gasses like CO_2, CH_4 (methane), and NO_x (nitrogen oxides) in real time across land, sea, and air in addition to allowing for sharper, faster interpretation of satellite data (Ghassemi

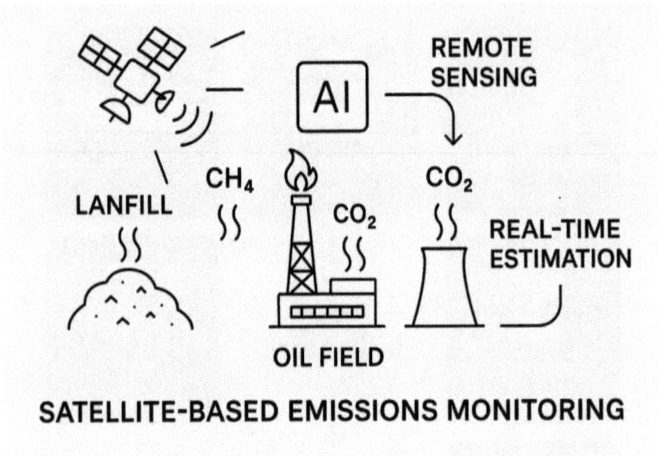

Fig. 14.2 Showcasing how satellites, enhanced by AI-driven remote sensing, are revolutionizing emissions monitoring. From detecting methane leaks at landfills to tracking CO_2 from oil fields and power plants, these technologies enable real-time estimation of greenhouse gases across land, sea, and air—ushering in a new era of transparency, regulatory enforcement, and global climate accountability

Nejad et al., 2024). Globally, this change has significant ramifications for emissions transparency, regulatory enforcement, and climate accountability (See Fig. 14.2).

Pattern recognition (Pourroosrtaei Ardakani & Cheshmehzangi, 2023a), particularly the identification of thermal anomalies and spectral characteristics linked to emissions sources, is the first step in integrating AI with satellite data. The distinct heat patterns of gas flaring at oil and gas extraction sites or the thermal plumes emitted by coal and gas-fired power plants can be detected by machine learning models trained on known emission events. AI systems detect these anomalies in almost real time using data from satellites such as NASA's MODIS, ESA's Sentinel-3, and NOAA's VIIRS. They also estimate the intensity, frequency, and duration of emissions-producing events. Such data has been around for a while, but it was underutilized since it was buried in petabytes of unstructured pictures before AI. Convolutional neural networks (CNNs) and deep learning have made it possible to continually and methodically examine these thermal and multispectral signals.

Machine learning is being utilized to directly predict the amount and kind of emissions in addition to thermal detection. In order to measure the amounts of CO_2, CH_4, and NO_x in the atmosphere (Xi et al., 2018), AI models now combine satellite-based spectral data with atmospheric chemistry models. Detailed measurements of trace gases are provided by instruments like the European Space Agency's TROPOMI on Sentinel-5P, NASA's OCO-2 and OCO-3 (Orbiting Carbon Observatories) (Guo et al., 2024), and private platforms like GHGSat (Alberge, 2021). These enormous datasets are processed by AI to determine the origins of emissions, account for air interference, and differentiate between natural and man-made fluxes. For instance, methane leaks from a particular pipeline or landfill can be identified using machine learning to separate them from the noise of background emissions.

Connecting satellite-observed emissions to changes in land use, particularly surface mining, peatland burning, and deforestation, is a potent use of this technology. Time-series footage is analyzed by AI algorithms to monitor changes in soil moisture, vegetation cover, and heat output. The carbon releases from land-use conversion are then calculated by comparing these changes with emissions inventories. In tropical areas like Indonesia, Brazil, and Central Africa, where forest loss is a significant contributor to emissions but is frequently underreported in national inventories, this integration is especially important. Preventative action is made possible by AI models' ability to forecast potential future emission locations based on past land change trends.

The capacity to accurately track particular infrastructure sources in real time (Cheshmehzangi, 2025), such as power plants, refineries, and metropolitan areas, is equally revolutionary. AI systems evaluate satellite-captured emissions plumes, compare them to known facility locations, and compute emissions using operational metrics such as heat production, plume length, or air dispersion. This makes it possible to create urban-scale emissions maps in heavily inhabited or industrialized areas (Pourroostaei Ardakani & Cheshmehzangi, 2023b), which show the various ways that industry, energy, and transportation affect local air quality and global warming.

The practical implementation of this is demonstrated by a number of significant efforts. Al Gore and associates founded the global coalition Climate TRACE, which generates independent, facility-level emissions estimates for all nations and major polluting sectors using satellite data, AI, and machine learning. It provides previously unheard-of transparency and granularity by combining data from over 300 satellites and numerous sensors. Methane and CO_2 “super emitters"—facilities that produce abnormally high emissions—are the subject of the Carbon Mapper project (Lyon, 2016), which is run by NASA's Jet Propulsion Laboratory and the University of Arizona. Carbon Mapper helps governments and businesses make quick mitigation efforts by using satellite-based spectroscopy to detect and measure these hotspots.

Through the Sentinel series, the European Union's Copernicus program has made significant investments in open-access satellite data. The Copernicus Climate Change Service (C3S) is developing AI-based initiatives to assess land degradation, identify trends in emissions, and assist EU member states with their national reporting. By bridging the gap between observation and policy action, these projects are propelling a surge of innovation in AI-powered emissions sensing.

These systems collectively represent the forefront of a new paradigm for emissions accountability that observes, evaluates, and measures carbon from above rather than waiting for governments or businesses to self-report. By transforming satellites into climate truth sensors, AI is pulling emissions out of the shadows and into the realm of verifiable, real-time public knowledge. By doing this, it is bringing unseen pollution into the open and giving us the means to address it.

14.3 AI in Supply Chain and Sectoral Carbon Footprinting

Emissions embedded in supply chains are a challenging and frequently opaque area of climate accountability that is receiving more attention as global attempts to achieve net-zero accelerate. Although they are the hardest to monitor, these so-called Scope 3 emissions—indirect emissions that happen across a company's value chain, from raw material extraction to product usage and disposal—often account for the biggest portion of a company's carbon footprint. The speed, complexity, and dynamic of international production networks are too much for traditional lifecycle analysis (LCA) and emissions accounting techniques to handle. AI is proving crucial in this regard, not just for bringing emissions into the public eye but also for incorporating them with previously unheard-of depth and clarity into daily business choices.

AI-powered lifecycle analysis dynamically tracks emissions across a product or service's entire life, going beyond static reporting. Large datasets from procurement systems, material inventories, energy consumption, logistics platforms, and financial records are fed into machine learning models to assess carbon footprints. By automatically classifying thousands of parts or transactions according to material type, location, mode of transportation, or production process, these systems may associate them with relevant emissions factors. AI allows businesses to compute emissions with site-specific accuracy and update them continually as suppliers or inputs change, eliminating the need for broad emissions averages.

Input–output modelling, a crucial technique made possible by AI, charts economic transactions across industries to determine how emissions affect the economy as a whole. Although input–output models have been around for decades, their use in carbon accounting has been constrained by computational complexity and data gaps. AI enhances both sides of the equation by employing statistical inference to fill in missing data and pattern recognition to fine-tune sectoral links. This makes it possible to estimate Scope 3 more precisely, particularly in industries where emissions are frequently hidden deep in upstream processes, such as electronics, food, and construction.

AI systems combine several data sources, including supplier statements, power bills, transportation logs, procurement records, and even financial movements, to create this all-encompassing image. Computer vision can scan packaging or barcodes for embedded emissions data, while natural language processing (NLP) technologies can extract pertinent information from contracts, invoices, and certifications. Together, these AI technologies produce comprehensive carbon maps of business processes, including the product SKU and delivery path. Finding carbon-intensive hotspots that contribute disproportionately to climate change, such as a single ingredient in a food product or a high-emitting supplier in an apparel chain, requires this degree of granularity.

AI also aids in the detection of emissions leakage, which occurs when supply chain emissions decrease in one area while increasing in another. For instance, if the supplier is farther away, moving to a low-carbon material can result in higher transportation-related emissions. Alternatively, an apparent drop in a company's reported emissions could be the result of outsourcing rather than true decarbonization. These changes can be identified and their net climate impact quantified by AI models trained on multi-sector data, guaranteeing that carbon accounting represents reality rather than optics.

AI is already being used in a number of businesses to measure carbon emissions. AI-powered platforms are being used by consumer products businesses such as Unilever and Nestlé to monitor emissions from ingredients, packaging, and delivery systems. These technologies assist them in determining which suppliers or goods have the biggest carbon footprints and in directing procurement or reformulation adjustments that lessen the impact. AI models in construction evaluate building materials like steel, cement, and insulation and monitor embedded emissions over the course of a project, allowing developers to select less impactful options and satisfy green building certification standards.

AI is enabling real-time optimization in global shipping and logistics, where emissions are influenced by a complex interaction of routes, vessel types, fuel choices, and cargo composition. In order to suggest lower-emission routes, maximize fuel efficiency, and flag inefficient or high-emitting boats, platforms such as Searoutes and RightShip employ AI to evaluate historical shipping data and meteorological circumstances. Additionally, AI is assisting ports and logistics centers in planning infrastructure enhancements that support decarbonization objectives and modeling emissions scenarios.

When taken as a whole, these advancements mark a paradigm change in emissions tracking. Carbon accounting becomes a dynamic system that is integrated into investment decisions, product design, logistics, and procurement, rather than being an after-the-fact sustainability reporting exercise. By turning emissions data from a compliance burden into a strategic asset, AI empowers businesses to take proactive, open, and efficient action. AI enables companies to cut costs across industries, borders, and supply chains by recognizing the entire climate cost of the things we purchase, construct, and export. This kind of visibility is not just helpful, but revolutionary in a world where carbon is limited.

14.4 Toward Radical Transparency: Policy, Ethics, and Planetary Accountability

AI is changing not just how we measure emissions but also how we regulate them as it becomes a crucial component of carbon accounting. AI is becoming the underlying infrastructure for radical openness at a time when credibility, clarity, and consistency are critical to climate action. AI is helping to standardize, validate, and expedite the collection, interpretation, and action of emissions data, from assisting business ESG disclosures and climate risk assessments to coordinating public policy with green taxonomies and international treaties. However, this revolutionary potential is accompanied by an urgent need to address the procedural, political, and ethical aspects of this expanding climate surveillance apparatus.

Governments and financial organizations are increasingly requiring companies and investors to make thorough carbon declarations. These disclosures are made possible by AI, which also makes them more thorough and comparable. By automating data collection and emissions calculations across Scopes 1, 2, and 3, it assists businesses in adhering to changing ESG frameworks, such as the EU's Corporate Sustainability Reporting Directive (CSRD) (Martinčević et al., 2024) or the Task Force on Climate-Related Financial Disclosures (TCFD) (Xiondole et al., 2025). AI offers the analytical tool required to assess if investments are in line with climate mitigation objectives for financial institutions governed by the EU's green taxonomy or the SEC's planned climate disclosure regulations. In this sense, AI promotes a more comprehensive legislative framework for the low-carbon transition in addition to voluntary sustainability initiatives.

However, there are also serious ethical concerns with AI's ability to make emissions obvious. The distinction between openness and surveillance is becoming increasingly hazy as AI-enabled monitoring grows, from satellites monitoring pollutants from factories to algorithms analyzing corporate buying records. Who gets to keep an eye on whom? Emissions data can be used as a tool for control when it is used for oversight. This is particularly delicate when tracking emissions in the context of Indigenous territory, sovereign nations, or informal economies. Accountability is encouraged by accessible emissions data, but it must also respect data sovereignty and be controlled by equitable, participatory mechanisms that shield marginalized populations from abuse or deception (See Fig. 14.3).

The conflict between proprietary and open access systems is the same. Many of the most potent AI technologies for measuring emissions are created by commercial businesses, frequently with funding from public–private partnerships. Despite their technological sophistication, these instruments might function as "black boxes," protecting their assumptions, algorithms, and data sources with intellectual property laws. Because only the most well-resourced actors may obtain or validate emissions data, this runs the risk of escalating existing imbalances in climate governance. Emissions claims are susceptible to manipulation or misunderstanding in the absence of consistent criteria for transparency, methodological disclosure, and third-party verification. This erodes confidence in climate reporting and impedes genuine progress.

The creation of open, interoperable frameworks for emissions monitoring must be given top priority by the international community to prevent this. Transparent, peer-reviewed models that are auditable and adaptable to different sectors and geographical areas should serve as the foundation for AI technologies. To guarantee that AI-driven carbon accounting is both thorough and inclusive, standardized emissions factors, verifiable data inputs, and publicly accessible algorithms are essential. Climate action necessitates uniform guidelines for measuring and confirming carbon that are based

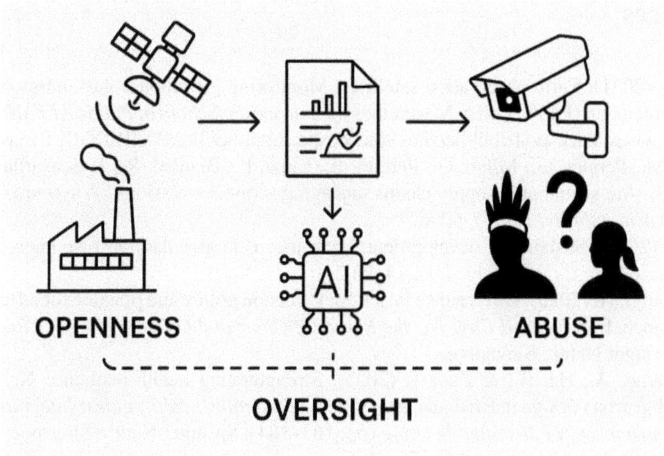

Fig. 14.3 This diagram captures the spectrum of AI use in emissions monitoring, ranging from transparency that drives accountability to the risk of surveillance overreach. It underscores the delicate balance between openness and oversight, highlighting how misuse of such technologies can harm Indigenous sovereignty, exploit marginalized communities, and transform vital environmental data into tools of control rather than empowerment

on science, regulated, and informed by equity, much like financial markets rely on common accounting standards.

Finally, AI's contribution to climate accountability extends much beyond efficiency or automation. Fundamentally, it is a climate truth-telling mechanism. It exposes pollutants that are hidden from view, whether they are strewn over a worldwide supply chain, buried in a subcontractor's operations, or integrated in a product. It opposes greenwashing by providing fact-based substitutes for exaggerated claims. Additionally, it democratizes knowledge access, allowing consumers and civil society to comprehend the actual carbon cost of our system in addition to corporations and regulators.

In this way, AI is a planetary mirror rather than only a carbon auditor. It enables us to see ourselves through the atmospheric reality we are creating together, rather than through spreadsheets or promises (Minevich, 2023). This awareness is crucial as we enter a century with a carbon constraint, not just to prevent disaster but also to create a future based on shared responsibility and quantifiable climate progress. When used responsibly, AI can serve as a link between knowledge and action, as well as between environmental risk and environmental stewardship. It gives us the ability to address the climate catastrophe with accuracy and purpose by giving the intangible tangible form.

References

Albergel, E. (2021). Earth observation satellites: Monitoring greenhouse gas emissions under the Paris Agreement. (Public Policy Master thesis). *SciencesPo School of Public AFFAIRS. En ligne:* https://www.sciencespo.fr/public/sites/sciencespo.fr.public/files/ALBERGELEve.pdf

Borchardt, M., Pereira, G., Milan, G., Pereira, E., Lima, L., Bianchi, R., & Scavarda do Carmo, A. (2025). Are sustainable supply chains managing scope 3 emissions? A systematic literature review. *Sustainability, 17*(13), 6066.

Brandon, P. (2012). Sustainable development: Ignorance is fatal–what don't we know? *Smart and Sustainable Built Environment, 1*(1), 14–28.

Cheshmehzangi, A. (2025). Governance in action: Precision policy and practice for advanced urban performance. In *Advanced City: For the Future of Cities and Opposing Smart Cities* (pp. 267–279). Springer Nature Singapore.

Cheshmehzangi, A., He, B., & Zuo, T. (2025). Strengthening health-resilience Nexus through sustainable urban design in Brisbane, Australia. In *Designing healthy cities: Integrating climate-resilient urbanism for sustainable living* (pp. 163–184). Springer Nature Singapore.

Ghassemi Nejad, J., Ju, M. S., Jo, J. H., Oh, K. H., Lee, Y. S., Lee, S. D., Kim, E. J., Roh, S., & Lee, H. G. (2024). Advances in methane emission estimation in livestock: A review of data collection methods, model development and the role of AI technologies. *Animals, 14*(3), 435.

Guo, K., Lei, L., Sheng, M., Ji, Z., & Song, H. (2024). Refining Spatial and temporal XCO2 characteristics observed by orbiting carbon observatory-2 and orbiting carbon observatory-3 using sentinel-5P tropospheric monitoring instrument NO2 observations in China. *Remote Sensing, 16*(13), 2456.

Küfeoğlu, S. (2024). Industrial process emissions. In *Net Zero: Decarbonizing the Global Economies* (pp. 341–414). Springer Nature Switzerland.

Lyon, D. R. (2016). *Quantifying, assessing, and mitigating methane emissions from super-emitters in the oil and gas supply chain.* University of Arkansas.

Ma, P., Dai, X., Guo, Z., Wei, C., & Ma, W. (2017). Detection of thermal pollution from power plants on China's eastern coast using remote sensing data. *Stochastic Environmental Research and Risk Assessment, 31*(8), 1957–1975.

Martinčević, I., Primorac, D., & Dorić, B. (2024). Corporate sustainability reporting directive (CSRD): Obligations, challenges and requirements for companies. *ENTRENOVA-ENTerprise REsearch InNOVAtion, 10*(1), 317–327.

Minevich, M. (2023). *Our planet powered by AI: How we use artificial intelligence to create a sustainable future for humanity.* John Wiley & Sons.

Pourroostaei Ardakani, S., & Cheshmehzangi, A. (2023a). Data-driven multi-target prediction analysis for driving pattern recognition: A machine learning approach to enhance prediction accuracy. In *Big data analytics for smart transport and healthcare systems* (pp. 63–80). Springer Nature Singapore.

Pourroostaei Ardakani, S., & Cheshmehzangi, A. (2023b). The role of big data analytics in urban systems: Review and prospect for smart transport and healthcare systems. *Big Data Analytics for Smart Transport and Healthcare Systems*, 1–26.

Wieland, A., & Creutzig, F. (2025). Taking academic ownership of the supply chain emissions discourse. *Journal of Supply Chain Management, 61*(1), 3–13.

Xhindole, C., Tarquinio, L., & Sierra-García, L. (2025). Climate change and task force on climate-related financial disclosures (TCFD) reports. A comparison between Italy and Spain. *Journal of Accounting & Organizational Change, 21*(7), 1–28.

Xi, Y., Ottinger, N. A., & Liu, Z. G. (2018). The dynamics of methane and NO_x removal by a three-way catalyst. *SAE International Journal of Engines, 11*(6), 1331–1342.

Yona, L., Cashore, B., & Bradford, M. A. (2022). Factors influencing the development and implementation of national greenhouse gas inventory methodologies. *Policy Design and Practice, 5*(2), 197–225.

Chapter 15
Intelligent Capital and Decoding Green: AI for Climate-Responsive Investment and Green Finance Revolution

15.1 Finance Meets the Planet: The Promise and Pitfalls of Green Investment

The financial industry has experienced a subtle yet significant change in recent years. The planetary limits of the twenty-first century are gradually redefining investment, which is no longer only driven by traditional risk profiles and short-term returns. Social inequality, ecological loss, and climate change have become tangible threats and demands for investment. This development has sparked the growth of green bonds, environmental, social, and governance (ESG) investing, and climate finance instruments that aim to raise money for sustainability (Joshi et al., 2024). Yet despite the promise of these innovations, the green finance revolution remains riddled with pitfalls—chief among them, the challenge of knowing what is truly green, impactful, and resilient in a world of cascading environmental risk.

Trillions of dollars' worth of global assets are currently managed through ESG investment, and green bonds are increasingly a popular way to fund initiatives ranging from clean energy to sustainable transportation and climate adaption (Leung et al., 2023). Climate-aligned finance flows have been further encouraged by international accords like the Paris Accord and the Glasgow Climate Pact, and development banks and sovereign funds have committed to incorporating climate risk into their financial operations. However, despite its growing volume, climate financing still faces credibility issues. Institutions and geographical areas use a wide range of metrics to assess ESG performance. Reporting standards are often vague, non-binding, or voluntary, making it difficult to assess whether investments deliver measurable climate outcomes or merely signal virtue.

Greenwashing, a rising issue in both public and private markets, is made possible by this ambiguity. Businesses might designate projects or portfolios as "green" without revealing clear information about long-term sustainability, ecosystem impact, or carbon reduction. Investors must make decisions about how to deploy their money based on climate performance, but they frequently lack timely, accurate,

© The Author(s), under exclusive license to Springer Nature Singapore Pte Ltd. 2025
A. Cheshmehzangi, *AI Agency in Climate Emergency*, Urban Sustainability,
https://doi.org/10.1007/978-981-95-4192-8_15

or consistent data to support these decisions. This presents an asymmetric information challenge. Additionally, most financial models do not yet appropriately price climate risks including drought, heat stress, sea level rise, and supply chain disruptions, which leads to blind spots in long-term risk management and asset pricing.

These difficulties highlight a significant discrepancy between the scope of the climate catastrophe and the instruments often employed to manage financial markets. They also highlight the pressing need for a new type of financial intelligence that is based on real-world environmental information and is dynamic and adaptive. This is where climate-responsive finance is greatly aided by AI. As such, it can provide transparency, verification, and predictive insight into green investment by combining large, complicated, and quickly evolving datasets, such as satellite imaging, emissions inventories, regulatory filings, and market activity.

News feeds, ESG reports, and financial disclosures can all be parsed by machine learning algorithms to find hidden climate risk exposures or identify discrepancies. Tools for natural language processing (NLP) can extract useful sustainability data (Ong et al., 2025) from unstructured text or look for greenwashing trends in business communications (Culpi Mann, 2024). By combining climate projection models with geographic data, deep learning models may assess the climatic risk of real estate or infrastructure assets (Wali & Bulla, 2024). Investors may now evaluate risk and impact at the asset level and in context with these tools, going beyond general ESG scores (See Fig. 15.1).

Furthermore, AI makes it possible to track ongoing monitoring, determining whether a project was "green" at first but also whether it produces long-term advantages. This real-time oversight is particularly important in high-risk or volatile contexts when transparency is poor and conditions on the ground change rapidly. By using crowdsourcing intelligence, remote sensing inputs, and proxy indicators to evaluate the actual climate performance of investments, AI can assist close the information gap in emerging markets and frontier economies where data may be limited or unreliable.

However, AI's capacity for prediction may hold the greatest revolutionary promise. AI may assist investors in projecting how their portfolios will perform in different futures by modelling scenarios of climate stress, policy change, and technology adoption. This allows for both risk reduction and opportunity discovery. As a result, finance's role is changed from reactive to anticipatory, proactively reallocating capital to industries, geographical areas, and innovations that promote long-term resilience and decarbonization rather than merely reacting to climate shocks.

Capital must aim for alignment with planetary boundaries in the climate period, not just development. That calls for a financial system that is as perceptive and adaptable as the dangers it encounters. AI is an essential partner in creating a climate-smart financial infrastructure, but it cannot take the role of judgment, ethics, or regulation. The current problem is to make sure that this intelligence is applied to alter the system rather than to game it; to reveal what is hidden, fix what is deceptive, and expedite what is genuinely sustainable.

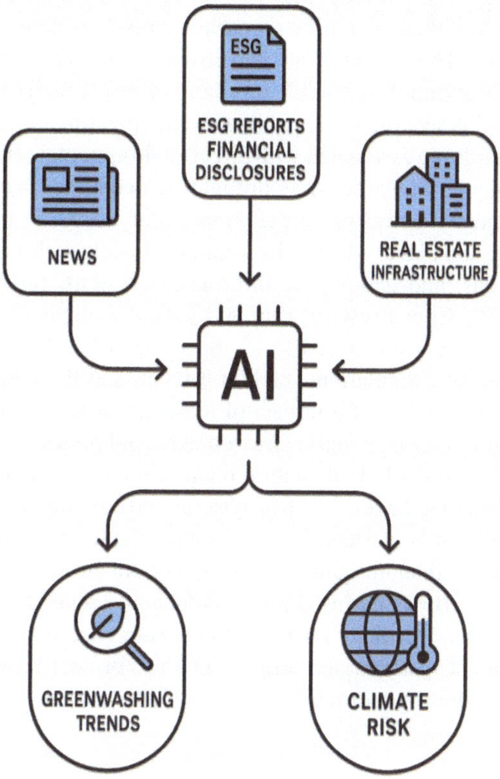

Fig. 15.1 This diagram shows how AI, powered by NLP and deep learning, analyzes news, ESG reports, financial disclosures, and asset data to expose climate risks and greenwashing. By cutting through corporate spin and sustainability fluff, these tools uncover discrepancies, false claims, and hidden exposures. They empower investors and regulators to see beyond inflated ESG scores, ensuring accountability and driving genuine climate action instead of polished PR narratives

15.2 AI for Investment Analysis and Adaptation Finance

Investment strategies must adapt to a new class of systemic threats as the climate catastrophe intensifies, moving beyond traditional risk-return models. In addition to being environmental issues, floods, fires, heat waves, sea level rise, and policy changes also have significant financial implications. Whether or not portfolios succeed in a warming environment depends on their ability to predict, measure, and react to these risks. This analytical gap is gradually being filled by AI, which gives governments, institutions, and investors strong tools for forecasting that takes climate risk into account (Bano, 2025) and for allocating funds to projects that are resilient, adaptive, and transformative.

By making risk visible, dynamic, and actionable, machine learning is transforming investment research that takes climate change into account. Conventional financial forecasting methods presume comparatively steady macroeconomic conditions and rely on historical data (Carriero et al., 2019). Climate change, however, challenges those presumptions. Contrarily, AI-driven models are able to learn to predict new dangers before they arise by ingesting complex datasets, such as weather forecasts, satellite images, policy scenarios, emissions patterns, and socioeconomic developments. These models give investors a more robust basis on which to make decisions by enabling the dynamic repricing of assets under various climatic futures.

AI is specifically capable of evaluating transitional and physical dangers. Physical hazards are associated with direct environmental effects, including floods in an industrial area along the coast, drought-related losses in agriculture, or severe heat-related damage to infrastructure (Forouheshfar, 2025). To measure these vulnerabilities over time and place, AI combines climate models with asset-level data, hydrological simulations, and geographic information systems (GIS) (Reece et al., 2024; Salunke, 2023). Conversely, market and policy changes linked to the shift to a low-carbon economy are referred to as transition risks. These include adjustments to energy subsidies, regulations, carbon pricing, and the uptake of new technologies. AI can assess the financial impact of these sector-wide transitions by analyzing regulatory changes, energy market movements, and policy pronouncements (See Fig. 15.2).

In the field of adaptation finance, which is the area of capital allocation aimed at enhancing the resilience of systems and communities to climate shocks, this dual-risk approach is particularly important. Adaptation, which has historically received less funding than mitigation, now offers both an economic opportunity and a moral obligation. By assessing the environmental, social, and financial feasibility of measures like heat-resilient infrastructure, water harvesting systems, coastal barriers, and drought-resistant agriculture, AI helps project-level adaptation funding. In addition to modelling return on adaptation (ROA) and assessing local susceptibility, machine

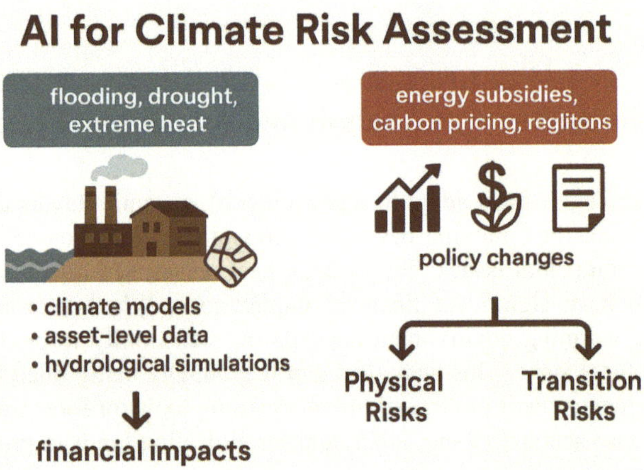

Fig. 15.2 AI evaluates climate risks by integrating climate models, GIS, asset data, and policy analysis. It assesses physical risks like flooding, drought, and extreme heat, alongside transition risks from energy subsidies, carbon pricing, and regulations, providing actionable insights into potential financial impacts

learning models also help quantify co-benefits like as avoided losses, employment creation, and biodiversity protection.

Applications of AI that are sector-specific are developing quickly. AI models in water infrastructure can identify regions most at risk of flooding or water scarcity by analyzing rainfall patterns, watershed dynamics, and urban hydrology. This information can then be used to guide investments in distributed rainwater systems, aquifer recharge, and wastewater recycling. AI in agriculture evaluates crop compatibility, soil health, and climate projections to guide investments in resilient seed systems, irrigation technology, and climate-smart farming. AI prioritizes cooling systems, green infrastructure, and neighbourhood-scale interventions for urban resilience by combining building data, heat stress mapping, and socioeconomic factors. Through these evaluations, financiers can more effectively match the priorities of climate adaptation and the Sustainable Development Goals with their portfolios (See Fig. 15.3.).

AI offers financial modelling for the transportation and sustainable energy industries that takes grid integration limitations, policy incentives, and climatic unpredictability into consideration. It simulates the lifespan costs and returns of electric transportation infrastructure, predicts the viability of wind and solar projects in climate-extreme regions, and calculates the future worth of energy efficiency retrofits under different regulatory frameworks. AI combines ecological, land-use,

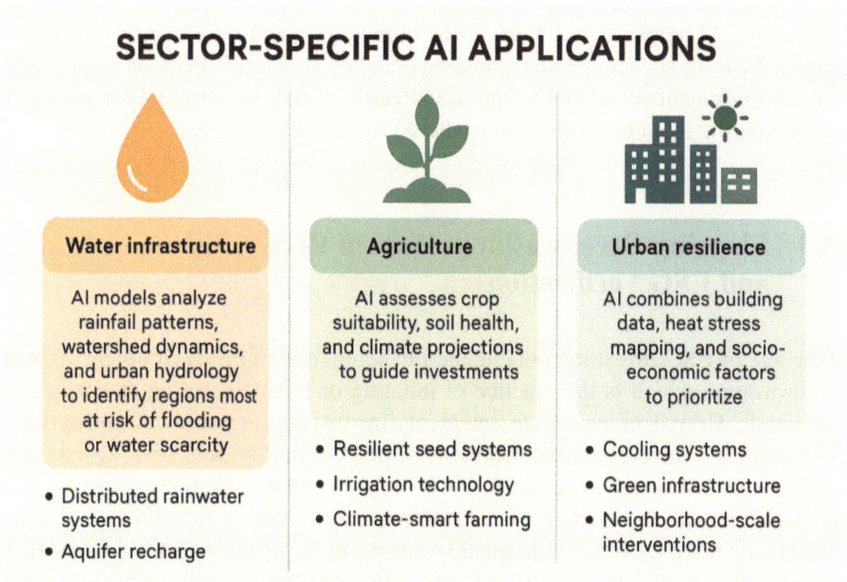

Fig. 15.3 AI drives climate adaptation by sector. In water infrastructure, it guides investments in rainwater systems and aquifer recharge. In agriculture, it supports resilient seeds, irrigation, and climate-smart farming. In urban resilience, it prioritizes cooling systems, green infrastructure, and neighbourhood-scale interventions for sustainable, targeted adaptation strategies

and market data to inform investments in regenerative land management, blue carbon ecosystems, and forest protection for nature-based solutions. These models aid in identifying the initiatives that succeed and those that fail in terms of quantifiable carbon, resilience, and community outcomes.

This move toward AI-powered climate finance is illustrated by a few practical examples. AI is being utilized more by climate tech venture capital firms like Lowercarbon Capital and Breakthrough Energy Ventures to assess ideas for system-level effect and climate scalability in addition to market potential (Croce & Bianchini, 2022). To assist governments in planning adaptive capital projects, development institutions such as the World Bank and Asian Development Bank are using AI into risk assessment platforms that measure climate vulnerability across infrastructure investments. Municipalities are employing AI to identify which urban enhancements, like solar rooftops, public transportation, and stormwater management, provide the best resilience returns for investors and communities in order to support green bond projects.

When taken as a whole, these applications suggest a new paradigm for climate finance: one that is performance-driven rather than merely green-labelled, and that is both predictive and reactive. In an age of growing uncertainty, AI adds precision, speed, and foresight to capital allocation. What are the climate-adjusted risks to my assets? is one of the important questions it helps investors answer. Where will the demand for adaptation come from next? Which investments provide equity, planetary value, and long-term resilience?

In short, AI helps finance to meet the climate moment with clarity, agility, and grounded intelligence rather than guessing or delay. This capability won't be optional as the requirement for global adaption increases. It will be essential for managing and prospering in an economy that is altered by climate change.

15.3 Flagging Greenwashing: Pattern Recognition and ESG Verification

In the age of ESG investing and climate financing, one of the most urgent issues is greenwashing, which is the practice of inflating or misrepresenting environmental credentials. The risk of inaccurate statements, inconsistent measures, and unverifiable disclosures has increased dramatically as trillions of dollars pour into "sustainable" portfolios and businesses scramble to fit in with green narratives. Trust in ESG claims is crucial in a market where sustainability is both a marketing tool and a category of material risk. AI is quickly becoming a potent watchdog, identifying discrepancies, confirming assertions, and enhancing the integrity of green finance, especially through developments in natural language processing (NLP) and cross-referenced analytics.

The capacity to parse language at scale is at the heart of this AI-driven oversight. Press releases, investor presentations, regulatory filings, and corporate sustainability reports are frequently full with story but lacking in factual detail. These writings can be scanned and interpreted by NLP algorithms, which can spot trends in internal contradictions, ambiguous promises, and overused keywords. An NLP model might, for example, highlight a business that makes a carbon neutrality claim in its investor update but leaves out explicit emissions information in its ESG report. It can identify inconsistencies between a declared commitment to supply chains free of deforestation and the ongoing inclusion of commodities sourced from at-risk sources in procurement agreements. Only human auditing would be able to identify such patterns, particularly when applied to tens of thousands of documents.

AI improves ESG verification beyond language analysis by cross-referencing reported data with data from third parties and actual performance. For example, comparing deforestation promises to remote sensing imagery, connecting corporate emissions claims to satellite-observed pollution, or comparing claims of renewable energy consumption with grid operator data. AI systems can cross-check corporate disclosures by ingesting data from social media, civil society monitoring networks, environmental NGOs, and public databases. AI can evaluate the veracity of a company's claim to use only renewable electricity by looking at previous energy procurement habits, facility locations, and the intensity of local grid emissions.

AI-driven grading and benchmarking systems that rank businesses' sustainability performance, climate policies, and exposure to ESG-related risks are also part of this capability. These systems analyze a variety of structured and unstructured data to produce impartial assessments; they don't just rely on company self-reporting. An AI tool might, for instance, model a company's decarbonization trajectory against sectoral climate benchmarks and established regulatory procedures to assess its transition strategy. It can assign a score to companies according to factors like regularity of third-party verification, transparency of Scope 3 emissions reporting, or alignment with science-based targets. Investors can then use these rankings to compare performance across industries or geographical areas, which aids in focusing funds on companies that are truly sustainable.

Importantly, AI also aids in the creation of early warning systems for regulatory and reputational risks in investments related to sustainability. AI systems can notify investors of developing issues long before they make the news by constantly searching the internet for updates on lawsuits, regulatory investigations, activist campaigns, or whistleblower stories. AI systems can identify these developments in real time, for example, if a bank is exposed for funding coal expansion despite green claims or a palm oil company is being investigated for illegal land clearing. This allows fund managers to evaluate exposure, modify positions, or have a conversation before reputational harm worsens.

This potential in international finance is already being shown by several AI platforms. AI is being used by startups such as Clarity AI, RepRisk, and Arabesque S-Ray to analyze ESG claims from thousands of businesses by analyzing millions of data points every day. For example, RepRisk creates dynamic risk profiles for businesses and industries by fusing natural language processing (NLP) with a well curated library of ESG disputes. It warns customers about hazards ranging from environmental carelessness to human rights abuses, including banks, insurance, and asset managers.

Impact verification is the main emphasis of Clarity AI, which compares business disclosures to industry benchmarks, local laws, and geospatial data. Major finance platforms use its tools to filter portfolios and make sure they meet sustainability standards.

AI is also being used by the insurance industry to evaluate resilience plans and climate claims. AI assists insurers in screening client disclosures on physical and transition risks as they deal with the growing losses from climate-related disasters, making sure that covered parties fulfill sustainability and resilience standards. AI is supporting credit risk models in banking that take ESG aspects into account, guaranteeing that climate-linked securities and green loans are supported by reliable, validated data.

AI is more than simply another data tool; it is a credibility engine in an investment environment where sustainability is both manipulable and mission-critical. It gives markets the ability to discern resilience from bluster and real impact from marketing jargon. AI will be essential to maintaining the credibility of green finance as it grows, making sure that the money flowing in the name of sustainability is not only well-meaning but also knowledgeable and well-founded.

15.4 Toward Climate-Conscious Capital Systems

Green finance must develop from a collection of promising mechanisms into a strong, reliable system for planetary transformation as climatic imperatives affect the global economy. The integrity of the mechanisms governing this evolution is just as important as the movement of capital. Whether through bonds, venture capital, insurance, or sovereign funds, green finance needs to be transparent, auditable, and grounded in ethical intelligence to have a significant impact. Long viewed as a tool for efficiency and speed, AI is now emerging as a key component of financial accountability, making visible what was previously hidden, verifiable what was presumed, and actionable what was previously too complicated to measure.

This revolution is built on transparency. With AI-driven tools that can track emissions, validate sustainability claims, and identify greenwashing, markets can finally start to match capital to climate reality. By requiring proof of effect, resilience, and planetary alignment rather than merely checking boxes on ESG disclosures, this change redefines what it means to invest responsibly. However, AI by itself cannot provide justice or fairness. Its algorithms must be open, its presumptions auditable, and its activities governed by moral standards. In an already unbalanced financial environment, AI could reinforce current injustices or turn into a tool of exclusion without these safeguards.

Increasing the availability of climate finance is one of the most pressing issues and opportunities. Developing countries, smallholder farmers, and Indigenous groups are frequently the most at risk from climate shocks, yet they are also the least able to obtain funds for adaptation or transition (Cheshmehzangi & Chen, 2021; Cheshmehzangi, 2025a; Keane et al., 2009). This disparity can be addressed with AI. AI can help underprivileged groups obtain vital funding by projecting local climate consequences (Ukoba et al., 2025), analyzing credit risk using non-traditional indicators, and streamlining compliance with green finance regulations (Siriman et al., 2025). AI makes climate funding more inclusive and participatory when combined with fintech innovations like digital ID systems, mobile banking, and blockchain-based disbursement platforms.

An essential component of this democratization is open finance data. It is imperative that governments, development banks, and charitable organizations take the lead in creating open-access platforms that enable the equitable application of AI tools across national boundaries, linguistic barriers, and financial systems (Allam & Cheshmehzangi, 2024; Cheshmehzangi, 2025b). Shared financial datasets have the potential to revolutionize climate investment, much like open climate models and satellite data have revolutionized environmental study. To enable civil society, researchers, and regulators to scrutinize money flows and hold actors responsible for fraudulent claims or misaligned portfolios, these platforms must be co-designed with transparency and equity in mind.

The use of AI in climate financing also necessitates careful policymaking. Regulation needs to balance innovation with oversight to ensure that AI technologies are fair, understandable, and devoid of manipulation. Digital finance strategies, green taxonomies, and climate-related financial disclosures must be incorporated into larger AI governance frameworks in order to ensure that the intelligence guiding capital allocation is both just and potent. To guarantee that climate-aligned AI is used responsibly and worldwide, not just in high-income markets, governments and international organizations must make investments in capacity building and cross-sector cooperation.

Lastly, this goes beyond markets. In an era of global danger, it is about creating a new framework of trust, wisdom, and shared accountability. AI is changing the foundation of climate finance, not just speeding it up. It enables us to act more confidently, see more clearly, and invest with greater purpose. By doing this, it turns climate finance into a vital component of resilience and regeneration rather than a specialized or compliance-driven industry. The direction of a liveable future as well as the performance of green portfolios will be determined by this intelligent capital system as the opportunity for action closes.

References

Allam, Z., & Cheshmehzangi, A. (2024). *Sustainable futures and green new deals* (pp. 1–137). Palgrave Macmillan.

Bano, S. (2025). The role of AI in predicting and mitigating climate change: A comparative study of US and China's approaches. *The Critical Review of Social Sciences Studies, 3*(1), 664–675.

Carriero, A., Galvao, A. B., & Kapetanios, G. (2019). A comprehensive evaluation of macroeconomic forecasting methods. *International Journal of Forecasting, 35*(4), 1226–1239.

Cheshmehzangi, A. (2025a). The importance of health-sustainability-peace Nexus in making Resilient Cities. *Peace and Sustainability*, 100012.

Cheshmehzangi, A. (2025b). Bridging innovation and sustainability: Global insights on ICT applications. In *Urban innovation and sustainability: ICT-based interventions in Hainan, China* (pp. 23–43). Springer Nature Singapore.

Cheshmehzangi, A., & Chen, H. (2021). *China's sustainability transitions: Low carbon and climate-resilient plan for carbon neutral 2060*. Springer Nature.

Croce, A., & Bianchini, R. (2022). The role of environmental policies in promoting venture capital investments in cleantech companies. *Review of Corporate Finance, 2*(3), 587–616.

Culpi Mann, E. (2024). *Enterprises' sustainability orientation: A natural language processing approach* (Doctoral dissertation). The University of Waikato.

Forouheshfar, Y., Ayadi, R., & Moghadas, O. (2025). Enhancing system resilience to climate change through artificial intelligence: A systematic literature review. *Frontiers in Climate, 7*, 1585331.

Joshi, D., Paul, S., Singh, P., Mane, N. S., & Yadav, M. (2024). Sustainable finance: The role of green bonds in driving ESG investments. *Library of Progress-Library Science, Information Technology & Computer*, *44*(3).

Keane, J., Page, S., & Kennan, J. (2009). *Climate change and developing country agriculture: An overview of expected impacts, adaptation and mitigation challenges, and funding requirements.*

Leung, H. H., Yorke, S. M., & Ng, A. (2023). 3 Social bonds as an innovative financial instrument for ESG initiatives. *Financial and Technological Innovation for Sustainability: Environmental, Social and Governance Performance.* Routledge.

Ong, K., Mao, R., Satapathy, R., Shirota Filho, R., Cambria, E., Sulaeman, J., & Mengaldo, G. (2025). Explainable natural language processing for corporate sustainability analysis. *Information Fusion, 115*, Article 102726.

Salunke, R. (2023). *Smart geotechnical asset management framework for climate adaptive infrastructure* (Doctoral dissertation). Jackson State University.

Reece, S., O'Donnell, E., Liu, F., Wolstenholme, J., Arriaga, F., Ascenzi, G., & Pywell, R. (2024). Assessing the Potential of AI for Spatially Sensitive Nature-Related Financial Risks. *arXiv preprint* arXiv:2404.17369.

Siriman, N., Ali, M. S., Siddiqui, T. A., Rohin, G., Vejju, B., & Cherupelly, N. (2025). Green finance and AI in India: A synergistic approach to sustainable development and climate resilience. *International Journal of Environmental Sciences, 11*(1), 207–216.

Ukoba, K., Onisuru, O. R., Jen, T. C., Madyira, D. M., & Olatunji, K. O. (2025). Predictive modeling of climate change impacts using Artificial Intelligence: A review for equitable governance and sustainable outcome. *Environmental Science and Pollution Research*, 1–20.

Wali, G., & Bulla, C. (2024). A data driven risk assessment in fractional investment in commercial real estate using deep learning model and fog computing infrastructure. *Library of Progress-Library Science, Information Technology & Computer*, *44*(3).

Chapter 16
Markets that Matter: AI for Carbon Credit Integrity and Climate-Aligned Trading

16.1 Carbon Markets at a Crossroads: Scale, Skepticism, and Opportunity

One of the most well-known strategies in the worldwide battle against climate change is the carbon market. They hope to generate financial incentives for decarbonization while allocating resources to mitigation and removal initiatives by placing a price on carbon emissions. Both the voluntary carbon markets, where businesses and people choose to offset emissions, and the compliance markets, which are required by law and linked to national emissions quotas, have grown rapidly in recent years. Carbon credits currently cover a wide range of initiatives, from direct air capture startups to methane capture in landfills to forest conservation in the Congo Basin. But even with their growth, carbon markets are at a vulnerable juncture, hampered by concerns about scale, integrity, and trust.

The caliber and legitimacy of offsets itself are among the main criticisms levelled at carbon markets (Blum, 2020). Concerns regarding "additionality"—whether the credited action would have happened even in the absence of market incentives—and "permanence"—whether the carbon collected or avoided would actually remain out of the atmosphere for decades or centuries—have become more widely expressed by the climate community. Public confidence has been damaged by reports of double counting, shoddy baselines, or non-existent projects. Watchdog organizations have found significant differences between reported and actual carbon reductions, and buyers frequently lack insight into what their money is truly supporting.

The ecosystem of the carbon market is also dispersed. There are numerous registries, standards, and techniques, and each has its own requirements for what constitutes a qualified project. Instead of providing freedom and innovation, this diversity has caused opacity and uncertainty. One registry may certify a project based on out-of-date information or lax verification, whereas another registry may reject the same project under stricter guidelines. The very idea of carbon trading as a trustworthy instrument for mitigating climate change is called into question by the

© The Author(s), under exclusive license to Springer Nature Singapore Pte Ltd. 2025
A. Cheshmehzangi, *AI Agency in Climate Emergency*, Urban Sustainability,
https://doi.org/10.1007/978-981-95-4192-8_16

lack of a standardized monitoring, reporting, and verification (MRV) system (Singh et al., 2016).

In this context, it is becoming increasingly clear that carbon markets require a digital update to operate efficiently or fairly. The pace, complexity, and scope of the modern carbon economy are too great for manual audits, sporadic site visits, and self-reported documentation. A new generation of tools that can increase openness, automate verification, and rebuild credibility is required. Here, AI is becoming a key enabler of reliable carbon market ecosystems rather than a panacea.

Real-time analysis of large, diverse data streams is made possible by AI. To determine whether a project is truly delivering on its promises in carbon markets, this entails combining satellite imagery, drone surveillance, Internet of Things sensors, environmental databases, weather records, and field reports. An AI model can, for example, track changes in tree canopy cover over time to verify forest offsets, identify illicit logging, or gauge the rate of reforestation. It may compare the claims made by a soil carbon project with crop data, hyperspectral soil images, and regional rainfall patterns. It can even identify questionable trends in the issuing of carbon credits that point to methodological manipulation, over-crediting, or double counting.

These AI-powered devices boost confidence in addition to accuracy. Trust is the foundation of markets, and at best, trust in carbon credits has been brittle. AI can assist in removing the information asymmetry that exists between buyers, sellers, and regulators by offering automated, auditable, and tamper-resistant verification methods. It transforms carbon accounting from a mysterious process to one that is open, constantly observed, and supported by data. Better price discovery and risk management are made possible for traders as a result. It improves compliance and supervision for governments and watchdogs. It makes finance more accessible and reduces transaction costs for project developers (Fig. 16.1).

AI's potential is far more expansive. By developing interoperable systems that translate and benchmark projects across several registries, it can aid in the harmonization of disparate standards. It can help with the development of dynamic crediting schemes that adapt to changing climatic baselines and real-time performance. Additionally, it can facilitate new kinds of climate-aligned trading, in which credits are regularly assessed for their effects on the environment, society, and economy rather than being certified only once.

Although carbon markets are fragile, they are not fundamentally flawed. If they lack integrity, their promise will be exposed. If they are done right, they have the potential to be one of the best tools for directing funds toward international climate action. AI adapts carbon markets to a world where emissions accountability needs to be quick, precise, and equitable—it doesn't solve them. AI will be essential to making sure that what is exchanged represents what benefits the environment as carbon markets approach a crucial decade of maturity.

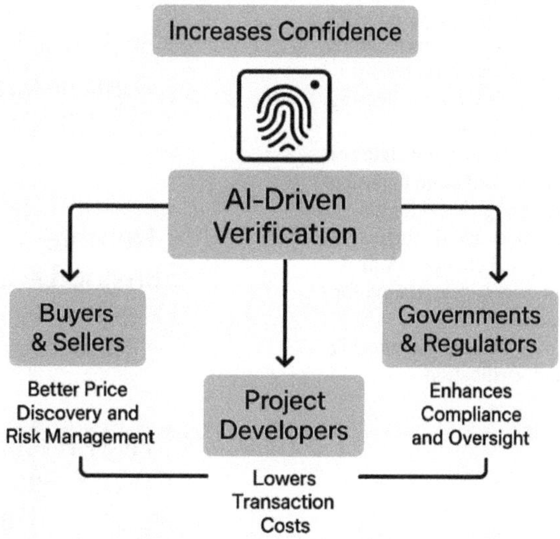

Fig. 16.1 AI-driven verification strengthens trust in carbon credit markets. By providing automated, auditable, and tamper-resistant systems, it benefits buyers and sellers through better price discovery, helps project developers lower transaction costs, and enables governments and regulators to enhance compliance and oversight, creating greater transparency and market confidence

16.2 AI for Offset Verification and Market Forecasting

Verifying carbon offset projects quickly, accurately, and in real time has become crucial as carbon markets grow to meet the demands of a world with a changing climate. With the ability to foresee market trends with a level of nuance that manual methods could never match, increase faith in the integrity of carbon credits, and enable more transparent and continuous monitoring, AI is entering this field with transformative promise. AI is becoming a crucial component of the contemporary carbon economy, from calculating credit prices across legislative changes to using satellites to confirm the efficacy of reforestation.

The integration of AI with satellite imagery and monitoring, reporting, and verification (MRV) systems is one of the most significant applications of AI in carbon markets (Islam, 2025; Sood, 2025). A worldwide, high-resolution picture of carbon sinks and emissions sources is provided by remote sensing data (Yin et al., 2021; Yu et al., 2022), which is obtained from satellite constellations like Landsat and Sentinel as well as from for-profit companies like Planet Labs. However, machine learning is needed to analyze such data at scale and transform it into insights that can be put to use (Chandra et al., 2022). To determine if a forest carbon project is actually increasing biomass or covertly decreasing it, AI systems can automatically identify changes in land cover, vegetation health, canopy density, and ground reflectance. These models are taught to identify small indications of illicit logging or fire damage that could compromise the integrity of the offset, or to differentiate seasonal variation from long-term degradation (see Fig. 16.2).

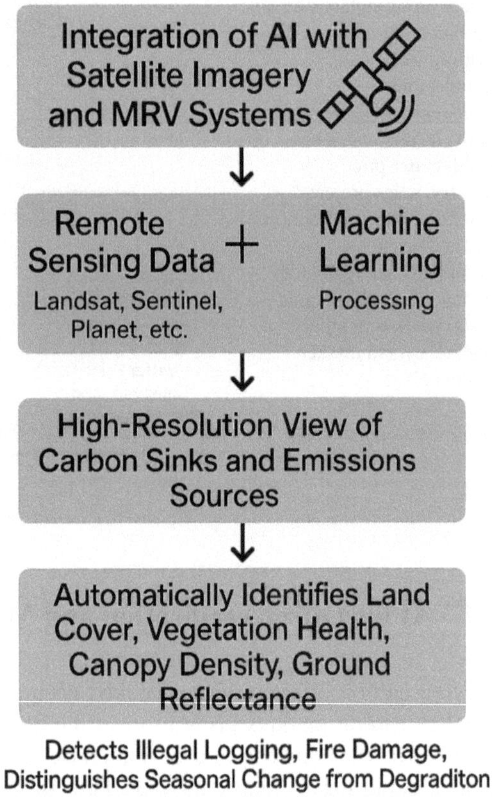

Fig. 16.2 AI integrates with satellite imagery and MRV systems to enhance carbon market integrity. By processing remote sensing data, AI delivers high-resolution insights on carbon sinks and emissions, detecting illegal logging, fire damage, and distinguishing seasonal changes from long-term degradation for accurate, real-time environmental monitoring and verification

AI models are currently being used to track soil carbon flux outside of forests by combining climate datasets, satellite data, ground-truth soil sampling, and hyperspectral imaging (Ding, 2024). AI-assisted models examine temperature data, plume patterns, and reported emissions in the context of methane capture projects, such as those in the energy, livestock, or landfill industries, to verify if capture devices are fulfilling their promises. AI can verify production and make sure credits are not overstated for renewable energy plants that are claiming offsets by comparing generation data with regional weather, national grid loads, and historical performance.

The use of self-reported documentation and sporadic field audits is lessened by this new paradigm of ongoing, AI-supported verification. Additionally, it adds more transparency and granularity, enabling purchasers and registrants to track the real carbon benefit achieved over time per credit. Businesses such as Pachama are already at the forefront of this movement (Loury, 2022; Pachama, 2025). Pachama provides offset buyers with real-time dashboards and dynamic forest carbon assessments by analyzing LiDAR data and satellite photos using deep learning. Similarly, Sylvera has developed a robust AI-based rating system that scores offset projects across dimensions like additionality, permanence, co-benefits, and implementation risk (Comel, 2025; Gupta, 2024), providing institutional investors with greater decision-making confidence.

Simultaneously, AI is showing great promise in detecting sectoral changes, modelling demand–supply dynamics, and predicting credit prices. These prediction models analyze a variety of factors, such as weather patterns, policy changes, market mood, and technological costs, and produce insights into the direction of markets. Machine learning models, for example, can predict which offset categories (such as nature-based, engineered removal, and energy efficiency) will see the highest volume growth in the upcoming years, model how changes in EU carbon regulations might affect global offset prices (Gasparini et al., 2024), or predict credit demand surges based on corporate net-zero commitments.

This forecasting skill is essential for improving liquidity, lowering volatility, and guiding long-term investment plans. AI models can be used by governments, traders, and climate funds to price carbon-linked derivatives, evaluate portfolio exposure, and schedule credit issuance according to anticipated supply restrictions. By providing probabilistic forecasts under various climate policies, energy transition routes, or technological adoption curves, AI models also assist scenario planning.

Crucially, AI assists in identifying hazards in addition to highlighting opportunities. AI systems can detect indications of project underperformance or manipulation by combining data from registries, field reports, news sources, and market movements (Corbet & Larkin, 2023). Models can identify geographic overlaps that may indicate double-counting, spot irregularities in credit issuance rates, or keep an eye on register transactions for odd activity. An offset project can be reported for investigation if, for instance, it often awards credits at higher-than-expected volumes or displays land-use data that is inconsistent with its approach. These early warning systems promote industry best practices while guarding against financial and reputational consequences.

AI-powered credit risk assessment systems are also growing in popularity (Cheshmehzangi, 2026). These technologies give projects and portfolios dynamic risk profiles that account for implementation, legal, and geopolitical risks in addition to climate variables. For instance, a forestry project in a politically insecure area would receive a lower permanence rating than one in an area with long-standing community stewardship agreements and good governance. Regulators can maintain environmental integrity and investors can more precisely price risk with this level of detail.

When taken as a whole, these AI applications signify a significant advancement in the functioning of carbon markets. They substitute clarity for opacity, real-time feedback for lagging measures, and informed foresight for conjecture. By doing this, AI redefines the legality of carbon trading in addition to improving its mechanics. Carbon credits must have unquestionable legitimacy at a time when they are anticipated to play a significant part in global decarbonization routes. AI can guarantee that each traded tonne of carbon reflects actual, quantifiable, and verifiable climate action if it is used wisely and inclusively.

16.3 Guarding the Ledger: Fraud Detection, Transparency, and Compliance

The complexity—and susceptibility—of carbon markets' operational ecosystems increases with their growth and maturity. The risks of fraud, manipulation, and regulatory non-compliance are present and growing as billions of dollars move through registries, exchanges, and over-the-counter transactions. Credible offset projects are not the only prerequisite for the validity of carbon markets; the integrity of the entire transactional infrastructure is also essential. Here, AI—which is already crucial for forecasting and verification—becomes even more important as the custodian of the carbon ledger, identifying anomalies, automating compliance, and enforcing transparency in carbon markets that are becoming more and more digitalized.

Credit duplication, in which the same emission reduction is counted more than once, either by separate businesses or across several market registries, is one of the most urgent issues in carbon trading. This may happen accidentally as a result of data inconsistencies or on purpose by taking advantage of registry fragmentation. In order to identify irregularities in issuance, ownership, or retirement, AI systems—especially those trained on extensive registry data—can analyze timestamped transactions, unique credit identifiers, and project geolocations. AI can quickly and accurately identify anomalies when a credit is given under several standards or shows up on various platforms without matching emissions reductions.

Duplication is not the end of manipulation. Inflated baselines, unsubstantiated effect claims, or speculative forward contracts that never materialize can also be the basis for credit issuance. AI algorithms can be trained to spot questionable trends in transaction flows or issuance, like sudden volume transfers to unconfirmed purchasers or a rise in project credits soon after a methodology change. By incorporating elements from the finance, cybersecurity, and logistics fields, these anomaly detection systems give carbon trading a much-needed extra layer of market monitoring.

Additionally, machine learning and natural language processing (NLP) improve the supervision of register records, legal contracts, and project documents. Project design documents (PDDs), emissions reduction plans, community benefit agreements, and verification reports can all be scanned by NLP techniques to determine credibility, identify discrepancies, and gauge standard compliance. NLP can identify a discrepancy, for example, if a REDD+ project in Central Africa asserts significant community involvement yet the documentation is devoid of participatory evidence. A trust measure can be added to a field that is frequently dominated by opacity by using AI techniques to provide reputation scores to developers or verifiers based on prior performance, audit results, litigation history, and public attitude.

In regulated carbon markets like the UK ETS, California's Cap-and-Trade Program, and the European Union Emissions Trading System (EU ETS), AI is also becoming increasingly essential to the automation of compliance. In these situations, companies are subject to stringent deadlines and criteria for reporting emissions, obtaining allowances, and retiring credits. By combining allowance tracking and reporting dashboards with emissions monitoring systems, AI applications streamline this procedure, minimize human error, and guarantee that compliance deadlines are fulfilled. AI guarantees consistent reporting and quick anomaly detection in high-volume industries with many assets across countries, such as utilities or cement.

Data normalization is one of AI's other important contributions. With dozens of registries, standards, and platforms employing various formats, metrics, and taxonomies, the carbon market landscape is infamously fragmented. Semantic harmonization-trained AI models are able to normalize inconsistent fields, translate between data formats, and produce unified dashboards that show real-time trading, retirement, and credit issuing activity across systems. For investors and regulators who want a unified perspective of market behaviour to evaluate climate performance, risk, and opportunity, this interoperability is essential.

The REDD+ (Reducing Emissions from Deforestation and Forest Degradation) programs throughout Africa provide a powerful illustration of this AI role. These initiatives, which frequently cover large areas with inadequate digital infrastructure, are vulnerable to problems with community trust and verification. In order to verify forest carbon claims, AI technologies now help by combining satellite photos, drone footage, local meteorological data, and project paperwork. AI has been used to identify unlawful encroachment, confirm forest regeneration, and highlight discrepancies between field reports and imaging in Ghana, Kenya, and the Congo Basin. By increasing openness and lowering the possibility of exploitation or erroneous crediting, these methods not only protect environmental outcomes but also boost community trust.

Similarly, AI-enabled credit validation systems that automate due diligence on small-scale clean cookstove or agroforestry projects are being piloted by fintech firms and policy organizations in southern Africa. By matching credits with registry records and scanning field data, GPS tags, and photographic evidence, these systems expedite the certification process and lower the possibility of "phantom" credits.

Ultimately, AI's contribution to the integrity of the carbon market goes beyond simply apprehending corrupt people; it also involves systematizing accountability. It substitutes dynamic, learning-based risk identification for static audits, spot checks for ongoing attention, and proactive oversight for ad hoc investigations. As carbon trading emerges as a key component of international climate policies, it needs to be safeguarded by intelligent mechanisms that maintain credibility, confidence, and influence at every stage of the market, not only by well-meaning individuals.

16.4 Designing Ethical, Inclusive, and Future-Proof Carbon Markets

The legitimacy of carbon markets will depend on trust rather than just innovation, scale, or money as they grow to meet the pressing need for global decarbonization. Furthermore, in the era of AI, trust must be gained through systems that are not just highly developed technologically but also morally sound, inclusive of all social groups, and openly managed. AI has the power to completely restructure carbon markets, turning them from disorganized, opaque transactions into quantifiable climate action engines. However, in the absence of intentional protections, it may potentially replicate and intensify the fundamental disparities and injustices that climate justice aims to eliminate.

The possibility of data asymmetry and surveillance overreach is one of the biggest dangers associated with AI-driven carbon credit schemes. Questions come up as AI systems absorb location data, contracts, sensor feeds, and satellite imagery: The data belongs to whom? Who determines its interpretation or the veracity of a project's carbon claims? We enter perilous ground when huge industrial companies face considerably fewer demands to demonstrate their creditworthiness than smallholder farmers, Indigenous land stewards, or community-based projects, who must submit to continuous digital surveillance. AI shouldn't be used as a weapon for unfair inspection, giving preference to those with more data while penalizing others with less.

Access equity is still a major challenge. Well-capitalized entities who can afford to pay experts, negotiate intricate certification procedures, and participate in protracted project development cycles have controlled many carbon markets. Meanwhile, smallholder or Indigenous-led ventures may face governmental, technical, and financial obstacles to entry. AI has the potential to either exacerbate or shrink this gap. AI may streamline MRV (Monitoring, Reporting, Verification) procedures, translate documents into many languages, offer mobile-accessible interfaces, and automate several time-consuming chores that presently exclude actors with limited resources if it is created with inclusivity in mind. However, this calls for more than simply technical optimization; it also calls for relationships with local communities, inclusive datasets, and intentional design (see Fig. 16.3).

The future of ethical carbon markets depends on community-integrated MRV models, decentralized registries, and open-source AI technologies (McKernon et al., 2024). Peer review, community supervision, and public confidence in procedures and decision-making processes are all made possible by open platforms. In addition to improving credit traceability and preventing manipulation, decentralized systems—based on distributed ledgers or blockchain—can enable local actors to co-govern the data ecosystems on which they depend. A way ahead that prioritizes lived knowledge in addition to digital intelligence is provided by community-led MRV models, where local stewards use AI to create, validate, and analyze environmental data.

Future carbon markets must be built with environmental, social, and technological resilience in mind. AI is used as a public utility to make sure that each carbon credit exchanged reflects an actual, additional, and equitable climate impact, not just to automate verification or forecast prices (Pourroostaei Ardakani & Cheshmehzangi, 2023). This necessitates norms for justice, transparency, and involvement in addition to accuracy and efficiency. From algorithmic training to data governance to the interface used for credit certification and trading, ethics must be ingrained at every level.

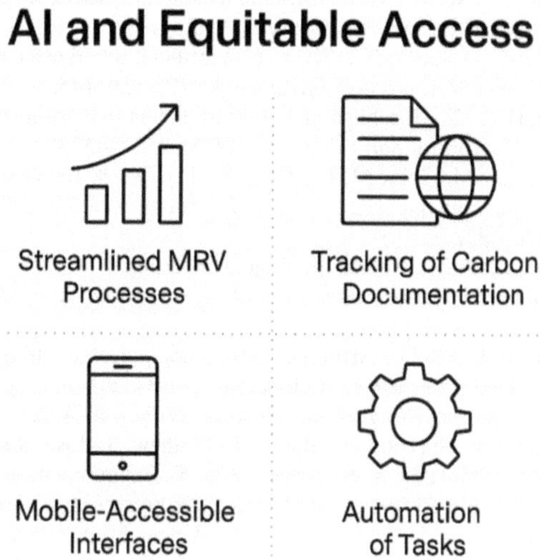

Fig. 16.3 AI can advance equitable access in carbon markets by streamlining MRV processes, translating documentation, enabling mobile-friendly platforms, and automating tasks—reducing barriers for smallholder and Indigenous-led climate projects

The ability of AI to ground the ethereal in verifiable reality is what really holds promise for carbon markets. When used properly, AI can reveal things that were previously invisible, such as the expansion of a forest canopy, the recharging of soil carbon, or the avoided emissions from a community microgrid (Joseph et al., 2025). It enables us to consider what really matters for the climate as well as for compliance. But only if AI systems are created with humility, shared governance, and a dedication to climate justice can that promise to come true. Carbon markets need to transform from flimsy systems of faith into strong systems of verification in this crucial decade. AI can serve as the framework for that development, but only if it is driven by the principles of equality, openness, and environmental stewardship rather than just the logic of efficiency. The way and for whom we choose to measure it will determine the future of carbon markets, in addition to what we can measure.

References

Blum, M. (2020). *The discursive legitimation of market-driven climate governance after the Paris agreement: The case of the global carbon offset markets* (Doctoral dissertation, Albert-Ludwigs-Universität Freiburg im Breisgau).

Chandra, S. N. R., Christopherson, J. B., Casey, K. A., Lawson, J., & Sampath, A. (2022). *2022 joint agency commercial imagery evaluation—Remote sensing satellite compendium* (No. 1500). US Geological Survey.

Cheshmehzangi, A. (2026). *Generative AI-powered urban digital twins: Pioneering environmental solutions for sustainable intelligent cities*. Springer.

Comel, L. (2025). *Scoring systems in the general data protection regulation and artificial intelligence act: A critical appraisal* (50 pp. document). Available online from https://www.maastrichtuniversity.nl/sites/default/files/2025-05/%283%29%20FINAL%20MCEL%20WPS%20Letizia_0.pdf

Corbet, S., & Larkin, C. (2023). Can artificial intelligence be used for target market manipulation. *Themes in Alternative Investments, 5*(2), 129–145.

Ding, H. (2024). Establishing a soil carbon flux monitoring system based on support vector machine and XGBoost. *Soft Computing, 28*(5), 4551–4574.

Gasparini, M., Ives, M. C., Carr, B., Fry, S., & Beinhocker, E. (2024). Model-based financial regulations impair the transition to net-zero carbon emissions. *Nature Climate Change, 14*(5), 476–481.

Gupta, K. (2024). Carbon credits and offsetting: Navigating legal frameworks, innovative solutions, and controversies. *International Research Journal of Multidisciplinary Technovation, 6*(2), 1–12.

Islam, F. S. (2025). Artificial intelligence-powered carbon market intelligence and blockchain-enabled governance for climate-responsive urban infrastructure in the global south. *Journal of Engineering Research and Reports, 27*(7), 440–472.

Joseph, L.P., Sherin, L., Ridha, Z.T., Mathew, S., Jose, S.K., Sebastian, T., Archana, L.S., Veena, R.K., Sreelatha, K. & Cyriac, J. (2025). Environment sustainability: Climate change mitigation, renewable resources, and biodiversity preservation aided by IoT and AI. In *The food-energy-water triangle for sustainable development* (pp. 269–291). Apple Academic Press.

Loury, H. (2022). Pachamama over people and profit: A case for Indigenous ecology and environmental personhood. *American Indian Law Review, 47*, 229.

McKernon, E., Glasser, G., Cheng, D., & Hadfield, G. (2024). AI model registries: A foundational tool for AI governance. *arXiv preprint* arXiv:2410.09645.

Pachama (2025). *See carbon in action.* Available online from https://pachama.com/

Pourroostaei Ardakani, S., & Cheshmehzangi, A. (2023). *Big data analytics for smart urban systems.* Springer.

Singh, N., Finnegan, J., Levin, K., Rich, D., Sotos, M., Tirpak, D., & Wood, D. (2016). *MRV 101: Understanding measurement, reporting, and verification of climate change mitigation* (pp. 4–5). World Resources Institute.

Sood, C. (2025). *From trees to tokens: AI and the digital renaissance of carbon markets.* Available at SSRN 5259696.

Yin, J., Fu, P., Hamm, N. A., Li, Z., You, N., He, Y., Cheshmehzangi, A., & Dong, J. (2021). Decision-level and feature-level integration of remote sensing and geospatial big data for urban land use mapping. *Remote Sensing, 13*(8), 1579.

Yu, Z., Deng, X., & Cheshmehzangi, A. (2022). The grain for green program enhanced synergies between ecosystem regulating services in Loess Plateau, China*Remote Sensing, 14*(23), 5940.

Part VI
AI for Climate-Conscious Citizenship and Behavioural Change

The potential of AI to enable people and communities to take significant climate action is examined in this section. Personalized eco-feedback tools help consumers develop more sustainable behaviours by guiding them in cutting back on waste and energy use. AI-powered platforms employ behavioural nudging and gamification—through simulations, messaging, and incentives—to promote climate-friendly decisions on a large scale. Additionally, by assessing public opinion, determining community objectives, and including a range of viewpoints into the policy-making process, AI improves civic engagement. AI acts as a catalyst for a more knowledgeable, involved, and environmentally sensitive society by coordinating individual actions with group objectives.

Chapter 17
From Awareness to Impact for the Reflective Citizen: AI-Powered Eco-Feedback for Sustainable Choices

17.1 From Information to Transformation: Why Feedback Matters

In the worldwide shift to a sustainable, low-carbon future, infrastructure, technology, and policy have received a lot of attention. Beneath all of this, however, is a more subdued revolution—one that is motivated by human behaviour. A critical change agent is the thoughtful citizen who understands how daily decisions affect the environment. However, awareness is rarely sufficient on its own. Feedback that is timely, pertinent, and empowering is essential for sustainable behaviours to establish and endure. Here, AI is starting to reshape how people relate to the climate crisis—not by bombarding them with facts, but by turning feedback into a customized catalyst for change.

The ability of feedback to form habits has long been acknowledged by behavioural science. Real-time cues promote action, such as the instantaneous effect of braking in a fuel-efficient vehicle or the decrease in electricity use after shutting off an appliance. Feedback generates a cycle of "see, act, adjust" that propels ongoing development. However, this loop is frequently interrupted for behaviours related to climate change. Static product labels, like "green" certificates or "energy efficient" ratings, don't offer any real-time advice. General lifestyle advice or quarterly energy bills are examples of delayed signals that don't effectively link action to impact. Static or generic messaging is no longer adequate in this age of climate crisis.

Immediate, interactive, and personalized tools are needed in light of the increasing pressure to minimize emissions and resource usage. Feedback systems must be able to adjust to people's habits, locations, preferences, and situations; they must be able to discern when to urge, when to nudge, and when to merely inform. Conventional methods are unable to generate the engagement and reflection required to promote significant change because of their one-size-fits-all design and slow response times. AI provides a path forward with its ability to learn, anticipate, and customize.

© The Author(s), under exclusive license to Springer Nature Singapore Pte Ltd. 2025

A. Cheshmehzangi, *AI Agency in Climate Emergency*, Urban Sustainability,
https://doi.org/10.1007/978-981-95-4192-8_17

Eco-feedback driven by AI is becoming a cutting-edge conduit between citizens and climate action. These systems convert complicated environmental data into intelligible, useful insights and are frequently included into wearable technology, smart meters, smartphone apps, and home energy dashboards (Cheshmehzangi et al., 2024). AI systems can contextualize data, such as by mapping usage to times of day, comparing it to similar households, or displaying equal carbon consequences, in place of displaying raw energy usage in kilowatt-hours. They can point out that last week's two-degree thermostat drop cut emissions by the same amount as planting a tree. These techniques bring the invisible to light by tying little choices to larger results (see Fig. 17.1).

AI-powered eco-feedback is unique in that it is both intelligent and flexible. To provide dynamic recommendations, machine learning algorithms can examine trends in home energy use, mobility preferences, water usage, and eating habits (Chen, 2024). At 10 PM, for instance, a user who frequently leaves appliances on standby would get a predictive popup asking whether they want to turn off their devices. On especially crowded or high-emission days, a commuter who frequently drives during

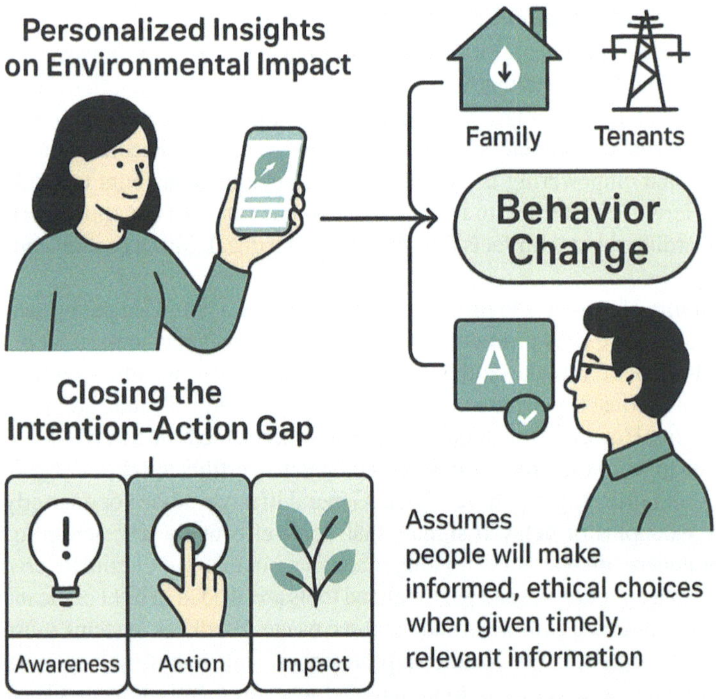

Fig. 17.1 AI-powered eco-feedback transforms climate stewardship into daily action. By providing personalized environmental insights, it fosters shared resource goals, improves building performance demands, and connects energy use to policy. Closing the intention–action gap, it empowers individuals and communities to make informed, ethical choices, turning passive awareness into sustained, impactful climate-friendly behaviours

rush hour can be encouraged to use public transportation, with specific advantages displayed. Subtle, anticipatory, and context-aware, these interventions seek to help rather than dictate; they are not pestering.

Crucially, AI makes it possible for feedback to be used for purposes other than monitoring. It becomes a tool for empowerment when it is created in an ethical and open manner. Eco-feedback systems can provide users with more agency, a better understanding of their influence, and more control over their consumption than surveillance (Snow et al., 2015). They can engage in gamified community challenges, compare themselves to their colleagues, set personal sustainability objectives, and monitor their progress over time (Saari, 2021). Feedback turns becomes a dialogue rather than an order—a mirror that encourages people to think and take action rather than a scorecard that places blame or shame on them (see Fig. 17.2).

The ramifications are significant. People start to integrate climate stewardship as a daily practice rather than a distant duty when they have access to personalized insights about their environmental impact. Families start working together to achieve

Fig. 17.2 AI-powered eco-feedback fosters empowerment through ethical, open design. It promotes user agency, gamified community challenges, peer collaboration, personal sustainability goals, and progress tracking—transforming feedback into a constructive dialogue rather than a tool of surveillance or control

common resource objectives. Better building performance is demanded by tenants. People start to understand how their consumption of power relates to more general energy policy (Burger et al., 2015). Once inactive or broken, the feedback loop is then closed, real-time, and full of potential.

Therefore, AI-powered eco-feedback is a key instrument for behaviour change and climate communication, not a supplementary convenience (Majid, 2024). Passive awareness is turned into significant, long-lasting influence, the abstract is operationalized, and the intention-action gap is closed. Its regard for the user's potential and intelligence, however, may be its greatest asset. It makes the assumption that individuals will act morally when they are provided with the appropriate information at the appropriate moment. This is not only hopeful, but necessary in a world that is on the verge of breakdown and transformation.

17.2 Designing Eco-Feedback Systems: Data, Personalization, and Engagement

Effective eco-feedback system design involves both data intelligence and behavioural involvement (Pourroostaei Ardakani & Cheshmehzangi, 2023). In the past, reports on energy or water use were frequently vague, slow, and unduly technical, giving people little idea of how they affected the environment or what they could do to improve it. AI is currently changing this environment by making dynamic, engaging, and tailored feedback systems possible. These systems do more than simply keep an eye on behaviour; they also react to it, learn from it, and encourage it to follow more enduring patterns.

Machine learning models that have been taught to identify and adjust to user behaviour, preferences, and environmental context are at the core of AI-powered eco-feedback systems. To create a profile of habits and tendencies, these systems continuously examine usage data, including trends in heating or cooling, appliance use, travel preferences, and even food consumption. They consider daylight hours, occupancy levels, local weather, and pricing during certain times of the day. By doing this, they improve their accuracy in providing fast, actionable, and pertinent feedback.

This intelligence is increased when combined with the burgeoning ecosystem of IoT devices, smart meters, and building management systems. For instance, a Google Nest smart thermostat learns when you are home, how you prefer your temperature at different times of the day, and then modifies the heating or cooling to suit your preferences. "Would you like to delay heating for 20 minutes to reduce emissions?" is a prompt that AI can display to customers after cross-referencing this data with the carbon intensity of the energy grid in real time. A straightforward comfort setting is transformed into a climate-conscious activity by this systemic convergence.

Similar to this, AI can dynamically coordinate lighting, HVAC, and ventilation in residential or commercial buildings with sophisticated building management systems

(BMS) based on environmental and occupancy data. Eco-feedback in this context could be suggestions to move operations to a more efficient window of time or alerts to building users when they've surpassed an energy benchmark. With an anticipated carbon savings, integrated platforms that link to mobility apps—like Citymapper or Google Maps—can even recommend more environmentally friendly modes of transportation, such recommending a shared bike for a nearby errand rather than a car (Fig. 17.3).

Disaggregation, or the capacity to divide total energy or water use into discrete parts, is a crucial aspect of these systems. By examining minute patterns in energy signatures, AI is quite good at this. The refrigerator's electricity consumption, the TV's duration of operation, and the washing machine's duration may all be seen in detail by users. With this level of detail, consumers can see exactly where they need to make changes and are no longer in the dark about their consumption.

The display of carbon footprints adds much more significance. AI systems convert kilowatt-hours or liters of water into measurable environmental measures rather than just reporting those numbers. Users can view how their carbon footprint compares to local or worldwide averages on a daily or weekly basis. These dashboards frequently offer recommendations, such as "Reducing your daily shower time by 15 minutes could lower your monthly carbon footprint by X kilograms of CO_2". By transforming

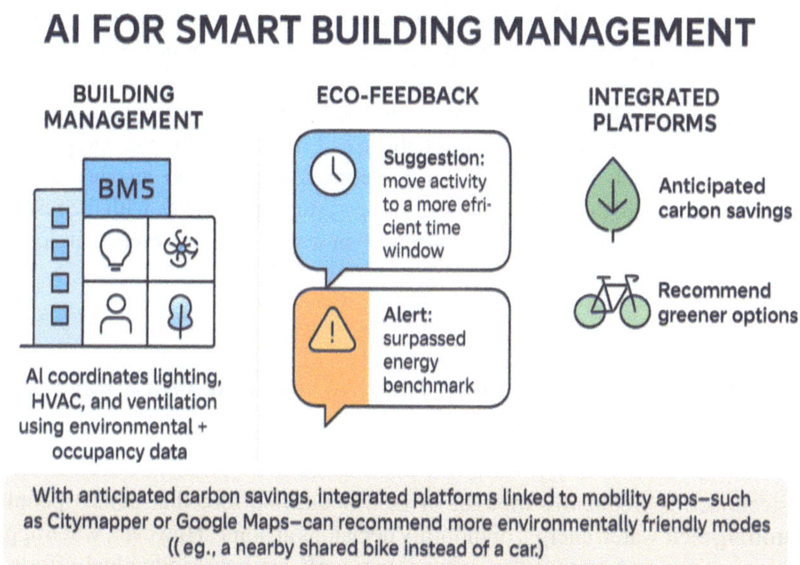

Fig. 17.3 AI-powered building management systems optimize lighting, HVAC, and ventilation using real-time environmental and occupancy data. Eco-feedback nudges users toward efficient usage and alerts them to overconsumption. Linked with mobility apps, these systems suggest greener travel modes, helping reduce carbon footprints and align daily operations with long-term sustainability goals

numbers into stories, this contextualization enables people to comprehend not only what they consume but also why it is important.

This is furthered by customized sustainability dashboards that are accessible via smart displays or mobile apps. They provide a customized experience where customers can establish objectives (such cutting emissions by 10% this month), monitor their progress, get personalized recommendations, and assess how they're doing in comparison to other users in their network. These dashboards frequently compile information from several sources, like as digital receipts, mobility trackers, and smart home devices, into a single, cohesive image of the impact.

Many systems now use behavioural nudging and gamification to increase participation. Community leaderboards, badges, and point systems encourage users to compete, work together, and get better (Bista et al., 2014). For example, consumers may receive incentives from nearby companies for reaching sustainability benchmarks or badges for continuously lowering energy consumption during peak hours. Some apps even set up "eco-challenges" in which homes or localities compete to cut waste or emissions the most over a predetermined amount of time.

These systems are additionally improved by behavioural nudging strategies that draw on behavioural economics. These could include social proofing ("80% of your neighbours used less water this week"), default settings that promote sustainable choices (e.g., switching to cold water wash cycles), or structuring communications to highlight both the financial and health benefits of environmental improvements (Potapov & Ramadan, 2025). The most effective nudges are ones that feel natural and encouraging, guiding users from intention to action without making them feel pressured or overwhelmed.

These AI-powered solutions aim to inspire new habits and ways of thinking rather than only alter behaviour once. Through the use of timely, intelligent, and personally meaningful feedback loops, they turn eco-consciousness from an abstract moral obligation into a real, everyday practice. These systems provide us with the resources to act sensibly—and the drive to persevere—in a world where every single action counts.

17.3 Beyond the Household: Institutional and Workplace Eco-Feedback

Eco-feedback is frequently thought of as a household tool that assists people in controlling their water, energy, or mobility decisions at home. However, when applied to organizations and workplaces, where teamwork can boost individual efforts and where culture change can have an impact on entire departments, teams, and even entire cities, the logic of feedback becomes much more potent. AI-powered eco-feedback is becoming a major force behind sustainable operations, behavioural change, and strategic decision-making in workplaces, public buildings, schools, and campuses.

AI solutions made for institutional environments may manage resource usage across entire networks as well as at the building level, providing real-time monitoring of transportation, waste, water, heating, and power, among other things. Environmental dashboards, procurement applications, and facilities management systems can all incorporate these elements. Beyond only measuring, they also forecast trends, learn from usage patterns, and provide practical insights that match day-to-day operations with long-term sustainability objectives. For instance, when energy consumption in a particular faculty building surpasses baselines, a university may receive notifications and AI-generated recommendations for lowering HVAC loads (Ahmad & Alshurideh, 2025) or improving lighting schedules (see Fig. 17.4).

Dashboards for organizations act as a communal feedback mirror. These dashboards display data that represents a team's or institution's environmental performance, split down by building, department, or activity, in contrast to personal apps designed for individual users. Accountability and cooperation are encouraged by this transparency. A department can start diagnosing the issue and working toward remedies when it discovers that it is using 20% more energy than its peers or that its emissions from travel are exceeding targets. For instance, procurement teams can identify high-emission vendors and recommend lower-impact substitutes by

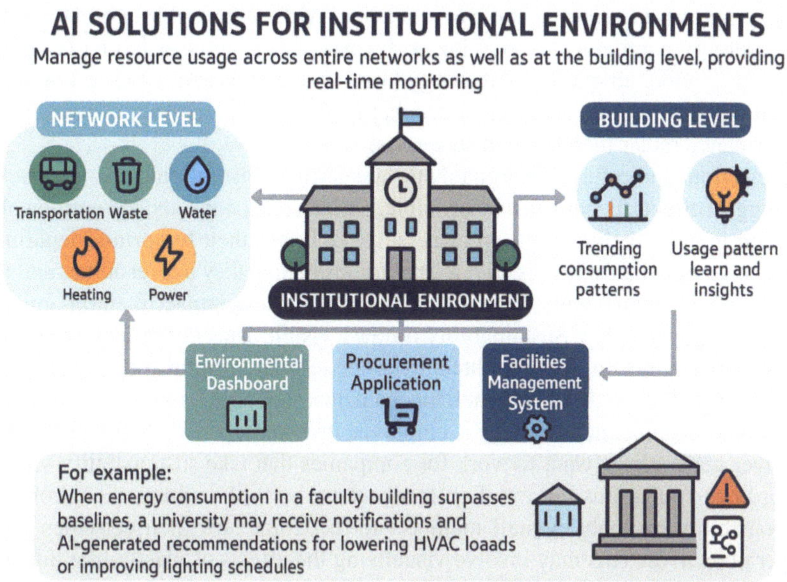

Fig. 17.4 AI solutions for institutional environments optimize resource management at both network and building levels, integrating transportation, waste, water, heating, and power monitoring. By learning from usage patterns, these systems provide actionable insights, real-time alerts, and sustainability recommendations, aligning day-to-day operations with long-term environmental goals while reducing inefficiencies and operational costs

using AI-integrated technologies to evaluate the environmental impact of suppliers or product categories.

It is particularly crucial to incorporate eco-feedback into procurement procedures. Purchased goods and services account for a large portion of an institution's environmental impact rather than direct operations (Buchanan et al., 2014). At the point of decision-making, AI technologies can review procurement data to assess embedded carbon, water use, or ethical sourcing criteria, producing sustainability scores. This puts environmental responsibility in the forefront and turns eco-friendly shopping into a habit rather than merely a desired policy.

More organizations are proving the worth of these platforms. AI-powered footprint tracking tools are currently being used by a few colleges in North America, Europe, and Australia to track emissions on campus, direct retrofitting plans, and facilitate sustainability reporting. These systems can model many scenarios, such as the effects of upgrading insulation, cutting back on air travel, or moving to renewable energy, and they may produce real-time visuals that assist leadership in setting priorities for interventions. In certain instances, the entire campus becomes a living sustainability laboratory when staff and students have access to this data via dashboards that are visible to the public.

Additionally, green certifications are changing. Institutions are starting to leverage real-time data streams to show compliance with certification standards rather than depending just on audits or static assessments. AI facilitates the automation of this procedure by continuously verifying performance indicators including trash diversion rates, indoor air quality, and energy intensity (Rautela et al., 2025). This lessens the administrative load and guarantees that green claims represent continuous, real performance rather than isolated successes.

Behaviour and culture are arguably the areas where institutional eco-feedback has the biggest effects. Feedback that is visible, regular, and connected to common objectives promotes a feeling of group ownership. To lessen their footprints, departments begin to compete. Employees start to spot trends in how they use resources and make the necessary adjustments. Teams that innovate in low-impact methods might be rewarded by leadership. Sustainability is now a visible aspect of the company's identity, debated in meetings, highlighted in newsletters, and demonstrated in day-to-day operations thanks to these feedback mechanisms.

Another significant advantage is employee involvement. Employees, particularly younger generations, want to work for companies that take sustainability seriously, according to numerous surveys. Eco-feedback tools build a culture of environmental responsibility by enabling staff members to recognize how their activities support larger objectives. This may involve visualizing the effects of plastic-free initiatives, measuring carbon savings from remote work policies, or even rewarding sustainable commuting through gamified platforms.

In conclusion, eco-feedback is no longer limited to personal apps or smart houses. It is turning into a tactical advantage in the institutional shift to sustainability. By combining the analytical power of AI with the behavioural insights of feedback loops, workplaces and educational institutions can evolve into engines of climate action—grounded not just in policy, but in data-driven, collective reflection.

17.4 Toward a Reflective Society: Equity, Privacy, and Climate Engagement

Eco-feedback systems driven by AI have the potential to revolutionize everyday life when they are incorporated into smart homes, college campuses, and other settings (Pittarello, 2016). However, the principles included into their design will also have an impact, in addition to technological advancement. We must address important issues of ethics, equity, and agency if we are to create a society that is really reflective—one that uses feedback not only to monitor but also to inspire, adjust, and empower. The social compact that regulates its usage will decide whether AI is an instrument of silent coercion or a freeing force, even though it may supply the intelligence for eco-feedback.

Data ethics is at the center of this conversation. Large amounts of environmental and personal data are gathered by sensors, smart meters, location monitoring, and digital interactions, and these data are the foundation of eco-feedback systems. This data is delicate and frequently discloses lifestyle, activity, and behaviour patterns that users might not be completely aware they are disclosing. Therefore, protecting privacy and openness cannot be compromised. Users need to understand what information is being gathered, how it will be used, who will have access to it, and how long it will be kept on file. Clear, revocable, and meaningful consent is required. User control should be given top priority in feedback systems, which should include options for anonymised data and fine-grained privacy settings without sacrificing system intelligence.

A fundamental component of any just eco-feedback system is equity, which goes beyond privacy. Individual residents, particularly those in underprivileged or disadvantaged communities, have little control over the infrastructure and policy decisions that underpin many of the most damaging activities in the globe. Eco-feedback technologies run the risk of escalating current disparities and shifting the burden of climate action to those with the fewest resources if they are only available to those with smart gadgets, dependable internet, or energy-efficient homes (Cheshmehzangi, 2020). Designing inclusive systems entails making them mobile-friendly, reasonably priced, and compatible with low-tech settings. It entails providing tools that function just as effectively in rural areas as they do in tech-savvy urban centers, bilingual interfaces, and culturally appropriate design.

It is imperative that eco-feedback never veers into surveillance. Helpful reminders and intrusive supervision are two sides of the same coin. Feedback shouldn't be intended to compel users to comply or to embarrass them; rather, it should promote introspection and voluntary behaviour change. Systems that respect human agency and offer choices and knowledge instead of strict regulations or harsh messages are the most successful. Instead than making people feel monitored or condemned, feedback should act as a mirror, not a microscope, giving them a better understanding of how they fit into the climatic landscape.

Participatory design methods can improve this regard for individuality. Users are more inclined to trust, interact with, and gain from tools when they are co-creators of those tools (Lundell et al., 2022). Communities may assist in determining what indicators are most important, what kind of input are most helpful, and what communication channels feel most empowering. This makes feedback a social practice that is based on discussion, experimentation, and a common goal rather than only a technical function.

The potential for AI-powered eco-feedback to raise awareness of climate change is more promising than its potential to merely optimize consumption or reduce emissions. The energy required for a morning shower, the carbon in a daily drive, and the emissions present in dinner are just a few examples of the effects of daily living that these technologies assist individuals in bringing to light. When these linkages become evident to us and we feel empowered to take action, we start to develop adaptive habits—not because we are required to, but because we want to.

Finally, kilowatts saved or emissions decreased alone won't be enough to gauge the effectiveness of eco-feedback systems. The degree of involvement they elicit, the inclusivity of their reach, and the caliber of awareness they foster will all be used to gauge their success (Ahmad & Islam, 2024). People who understand their position in the climate system—not as passive consumers, but as active contributors to a shared, low-carbon future—are living in a truly reflective society. AI can assist us in getting there—clearly, not with orders. by enhancing human judgment rather than taking its place. By assisting us in seeing ourselves, not by observing us.

References

Ahmad, A. A., & Alshurideh, M. T. (2025). Revolutionizing building management: The ORBIT FRAMEWORK for IoT-powered smart buildings. *International Journal of Management and Marketing Intelligence, 2*(2), 1–13.

Ahmad, I., & Islam, M. R. (2024). Empowerment and participation: Key strategies for inclusive development. *Building strong communities: Ethical approaches to inclusive development* (pp. 47–68). Emerald Publishing Limited.

Bista, S. K., Nepal, S., Paris, C., & Colineau, N. (2014). Gamification for online communities: A case study for delivering government services. *international Journal of Cooperative information Systems, 23*(02), 1441002.

Buchanan, K., Russo, R., & Anderson, B. (2014). Feeding back about eco-feedback: How do consumers use and respond to energy monitors? *Energy Policy, 73*, 138–146.

Burger, P., Bezençon, V., Bornemann, B., Brosch, T., Carabias-Hütter, V., Farsi, M., Hille, S. L., Moser, C., Ramseier, C., Samuel, R., & Sander, D. (2015). Advances in understanding energy consumption behavior and the governance of its change–outline of an integrated framework. *Frontiers in Energy Research, 3*, 29.

Chen, J. (2024). Building eco-feedback system and occupants' energy using behavior. *Circular economy for buildings and infrastructure: Principles, practices and future directions* (pp. 259–274). Springer International Publishing.

Cheshmehzangi, A. (2020). Evaluating the Nexus between housing and energy sectors: The comparison of urban—Peri-Urban and rural housing areas in Zhuhai, China. *Energy and Power Engineering, 12*(6), 314–338. https://doi.org/10.4236/epe.2020.126019

Cheshmehzangi, A., Batty, M., Allam, Z., & Jones, D. S. (2024). City information modelling: An insight into a new era for the built environment. In *City information modelling* (pp. 1–13). Springer Nature Singapore.

Lundell, S., Toots, A., Sönnerfors, P., Halvarsson, A., & Wadell, K. (2022). Participatory methods in a digital setting: Experiences from the co-creation of an eHealth tool for people with chronic obstructive pulmonary disease. *BMC Medical Informatics and Decision Making, 22*(1), 68.

Majid, G. M. (2024). *Nudging at scale: Uncovering conversational AI potentials to facilitate pro-environmental behaviour spillover* (Doctoral dissertation). University of Surrey.

Pittarello, F. (2016, July). From building control to eco-feedback: Opportunities and challenges. In *2016 Intl IEEE Conferences on Ubiquitous Intelligence & Computing, Advanced and Trusted Computing, Scalable Computing and Communications, Cloud and Big Data Computing, Internet of People, and Smart World Congress (UIC/ATC/ScalCom/CBDCom/IoP/SmartWorld)* (pp. 1013–1020). IEEE.

Potapov, K., & Ramadan, M. (2025). *Participatory design of an eco-feedback tool with teens for community energy futures.* Available at SSRN 5365671. https://papers.ssrn.com/sol3/papers.cfm?abstract_id=5365671

Pourroostaei Ardakani, S., & Cheshmehzangi, A. (2023). The role of big data analytics in urban systems: Review and prospect for smart transport and healthcare systems. *Big Data Analytics for Smart Transport and Healthcare Systems* (pp. 1–26). Springer.

Rautela, K. S., Goyal, M. K., & Surampalli, R. Y. (2025). AI and machine learning for optimizing waste management and reducing air pollution. *Journal of Hazardous, Toxic, and Radioactive Waste, 29*(3), 04025014.

Saari, A. (2021). *Changing behavior towards sustainable consumption habits with eco-feedback technologies through user experience.* Available online from https://aaltodoc.aalto.fi/items/2558f6e7-f20f-43b5-8e0e-86feefbc0eb0

Snow, S., Vyas, D., & Brereton, M. (2015). When an eco-feedback system joins the family. *Personal and Ubiquitous Computing, 19*(5), 929–940.

Chapter 18
Gamification and Incentivising Impact: AI, Nudging, and the Psychology of Sustainable Action

18.1 Behavioural Bottlenecks in Climate Action: The Gap Between Knowing and Doing

Climate change is a behavioural issue rather than just a technical or policy one. The majority of individuals find it difficult to convert their increased awareness of the environmental situation into consistent action (Cheshmehzangi & Dawodu, 2018). This is a reflection of the profound behavioural bottlenecks that characterize human decision-making, not a failure of intent. The frustrating gap between knowing and doing is caused by a combination of factors such as inertia, decision fatigue, ingrained behaviours, and the abstract nature of climate impacts. Even though our daily decisions—what we eat, how we travel, and what we purchase—have an impact on the carbon footprint of modern living, they are rarely done in logical, sequential ways. The chaotic, emotional, and routine-rich reality of everyday life serve as the backdrop for their creation.

One of the biggest barriers to behavioural change is complexity. Many climate-friendly practices, such as cutting back on energy consumption, changing means of transportation, or handling household trash, call for preparation, mental work, and lifestyle changes. Individuals are occupied. Even if they are aware that it is not sustainable, they fall back on what is comfortable. For this reason, even well-written information is rarely enough to spur change. Although carbon labels and educational initiatives can increase awareness, they frequently lack emotional connection, customization, and immediacy. When action has immediate costs or discomforts and distant, nebulous rewards, the gap grows even more. Recycling could seem tiresome. It could seem restrictive to avoid flying. The benefits of such actions—reduced emissions, ecosystem protection, climate stability—are distributed across space and time, often too abstract to motivate sustained shifts in behaviour.

Affective and experiential approaches that are based on people's true feelings and behaviours are needed to solve this. Interventions are more likely to be successful if they meet individuals where they are instead of where models suggest they should be.

© The Author(s), under exclusive license to Springer Nature Singapore Pte Ltd. 2025

A. Cheshmehzangi, *AI Agency in Climate Emergency*, Urban Sustainability,
https://doi.org/10.1007/978-981-95-4192-8_18

Here, AI has a potent and mainly unrealized potential: it can serve as a behavioural architect, creating context-sensitive settings that make the sustainable option the default, preferred, and easiest one. AI is able to deduce goals, learn from patterns in behaviour, and customize interventions for different situations. It can predict what types of feedback people respond to most effectively or when they are most likely to be open to a gentle prod.

The emphasis is shifted from merely supplying more information to creating better choice settings when AI is framed as a behavioural architect. It has to do with reorganizing the circumstances surrounding decision-making. Even minor adjustments to timing, presentation, or framing can have significant effects. Smartly timed and personally relevant micro-interventions, such as a default setting that puts an appliance in energy-saving mode, a reminder to take public transportation when the grid's carbon intensity is high, or a reminder to avoid meat on days when emissions are at their highest, can encourage people to make better decisions without taking away their freedom.

The term "nudging," which has its roots in behavioural economics, describes subtle design elements that influence behaviour while maintaining agency. AI-powered nudging improves its effectiveness, adaptability, and personalization. Based on factors like the time of day, past behaviour, or even the emotional tone of digital interactions, machine learning algorithms may determine when a person is most likely to respond to a recommendation and provide nudges at the ideal time. Crucially, these prods don't have to be explicitly moralistic. When people feel helpful, insightful, or even playful, they perform at their best.

This leads to gamification, which is the application of game characteristics outside of games to boost engagement and motivation (Alsawaier, 2018). Leaderboards, progress bars, badges, and points all appeal to our innate urge to succeed, compete, fit in, and get better. Gamification makes sustainability a living, interactive experience when it comes to climate action. AI can suggest challenges that are appropriate for each player's lifestyle, personalize game mechanics, or organize team competitions that make climate action more social and communal. To reduce their weekly energy consumption, a family may compete. Green commuting may offer incentives for office workers. Students might collaborate to reduce campus waste, with real-time dashboards showing collective impact (see Fig. 18.1).

When properly built, both gamification and nudging acknowledge a basic reality: behaviour change is about enabling, not convincing. It involves establishing settings, signals, and rewards that encourage people to behave in ways that are consistent with their best intentions. AI facilitates agency rather than acting as a change agent. It offers something considerably more important in a world when climate information is abundant: the appropriate trigger, at the right moment, in the right method to transform knowledge into action that has enduring impact.

Fig. 18.1 Gamification transforms climate action into an engaging, social challenge. AI personalizes goals, tracks progress, and fosters collaboration through leaderboards, badges, and points. From families competing to save energy to students reducing campus waste, every participant contributes to measurable, collective environmental impact in real time

18.2 Nudges that Learn: AI for Tailored Messaging and Predictive Habit Design

Static strategies are no longer adequate in the evolving field of climate involvement. While a sign reminding someone to switch off the lights or a label stating a product's carbon footprint are examples of traditional nudges, the potential and difficulties of today call for more clever, flexible solutions. Presenting AI-powered nudging, a novel approach where the nudge is responsive, dynamic, and customized to the person's changing circumstances (Mohammadian, 2024a, 2024b). These interventions are not universally applicable. They are learning nudges.

Machine learning models that have been trained to identify trends in user behaviour are at the core of this strategy. The time of day, location, weather, past decisions, lifestyle preferences, digital interactions, and even inferred goals are just a few of the many inputs that these models continuously consume and evaluate. The system builds a behavioural profile over time that makes it possible for it to provide nudges that are more timely, relevant, and effective. For instance, based on dietary requirements, previous purchases, and local availability (Ajayi, 2025), a person who typically shops for groceries on Saturday mornings would get a Friday night prompt with seasonal, low-carbon dish ideas (Kim, 2017). AI may encourage a user who usually drives to work but occasionally rides a bike on sunny days to do so when it predicts favourable weather and less traffic (see Fig. 18.2).

Customized interventions like this are not a coincidence. Goal-oriented recommendation engines that operate in the background to match individual incentives with global goals are the cause of them. These solutions, in contrast to strict behavioural programs, provide dynamic goal setting, enabling users to select their own climate goals, such as cutting back on electricity consumption, consuming less red meat, or avoiding single-use plastics. The AI adjusts the challenges, recommendations, and feedback based on the speed and response of each user. The system shifts to more manageable steps if a goal is often missed, rewarding progress over perfection. If a goal is reached with ease, the system makes it harder.

Timing is everything. The timing and manner of a message's delivery can have a greater impact than the content itself, as behavioural science has long demonstrated. AI is excellent at timing methods because it can analyze contextual indicators like real-time transport delays, grid carbon intensity, and ambient temperature in addition to user habits. When buses are on time, private vehicle emissions are high, and the user has no conflicts with their calendar, a push message may be issued to encourage the use of public transportation. The outcome seems more like a recommendation from a reliable advisor than a demand.

In many different fields, this dynamic nudging concept is used in practice. AI-powered smart thermostats and energy dashboards, for instance, monitor household trends and suggest behavioural changes in energy use, such as pre-cooling before peak hours or delaying appliance use until solar availability is high. Based on user location and environmental factors, ride-sharing applications like Uber and Lyft have started experimenting with AI prompts that recommend bike routes, walking routes,

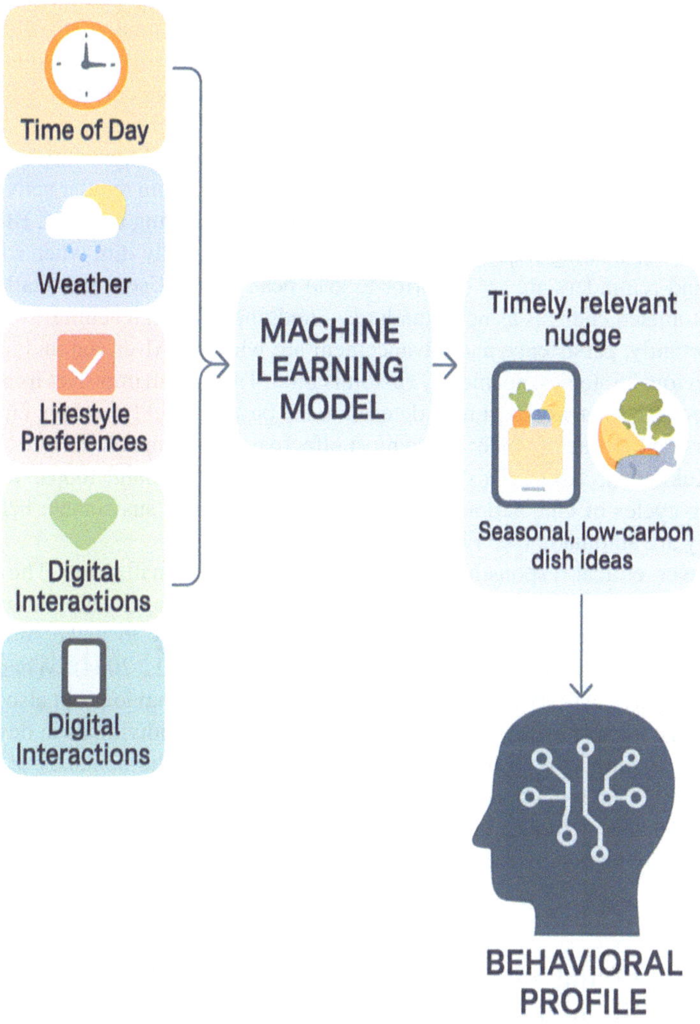

Fig. 18.2 Machine learning uses diverse inputs—such as time of day, weather, lifestyle preferences, and digital interactions—to create a behavioural profile. From this profile, AI delivers timely and relevant nudges, like seasonal low-carbon meal suggestions, tailored to individual habits and circumstances

or shared trips. In addition to lowering emissions, these nudges offer frictionless substitutes that are handy and consistent with user intent.

Food waste and dietary emissions are being addressed by nudges that AI-powered supermarket platforms are presently testing. The system may suggest reduced servings, meal plans that utilize leftovers, or recipe changes that substitute low-impact components (like lentils or mushrooms) for high-impact ones (like meat) based on its ability to identify what is in a user's cart and anticipate future spoilage or over-purchase trends. Similar approaches are being used by retailers to offer "green bundles", customized discounts on climate-friendly products, or sustainability scores based on past purchasing history. This effectively encourages consumers to choose less impactful options without calling for change.

Another frontier for predictive habit design is represented by smart homes. AI-powered gadgets, such as Google Nest or Alexa, can learn from regular activities and gently change settings to lessen impact without compromising comfort. Blinds can adapt to lower heating requirements, lights can automatically dim when rooms are empty, and reminders are set off prior to grid peak hours. Users eventually accept these automated changes as new standards—nudging without teaching.

Importantly, persistence and advancement are what set AI-driven nudging apart from previous strategies, not merely customisation. The system improves its approach as it discovers what works. It may determine the best channel (text, app, email), the most persuasive tone of voice, or the most effective peer comparison style (Lehto & Oinas-Kukkonen, 2011). This makes it possible to create habit loops, which are recurring cycles of cue, action, and reward that strengthen sustainable behaviours until they are automatic (see Fig. 18.3).

However, ethical responsibility accompanies such personalization. The purpose of nudging should continue to be empowerment rather than manipulation. Value alignment, transparency, and opt-out choices are necessary to make sure the AI acts more like a collaborator than a puppeteer (Adler et al., 2024). When implemented properly, AI-powered nudging not only changes behaviour but also lays the groundwork for self-awareness and adaptability, assisting individuals in developing routines that support both their individual objectives and the demands of a struggling planet. Our biggest collective lever in a world with climatic constraints is behaviour change. With AI-driven nudging, it becomes a scalpel—precise, flexible, and incredibly human-aware—instead of a blunt tool.

18.3 Gamifying Sustainability: Rewards, Challenges, and Simulation Worlds

One of the most promising strategies for transforming sustainability from a moral requirement into an interesting, interactive experience is gamification. Gamification reimagines climate action as something to strive for, enjoy, and share by appealing to our innate human motivations—achievement, competition, curiosity, and social belonging. It provides a bottom-up ecology of challenges, incentives, and fun experimentation as opposed to top-down directives or abstract warnings. It does not speak of guilt, but of motivation.

Fundamentally, gamification employs components like leaderboards, badges, points, and social comparison to encourage behaviour. You could receive points for finishing an energy-saving task. Badges are unlocked by completing several tasks.

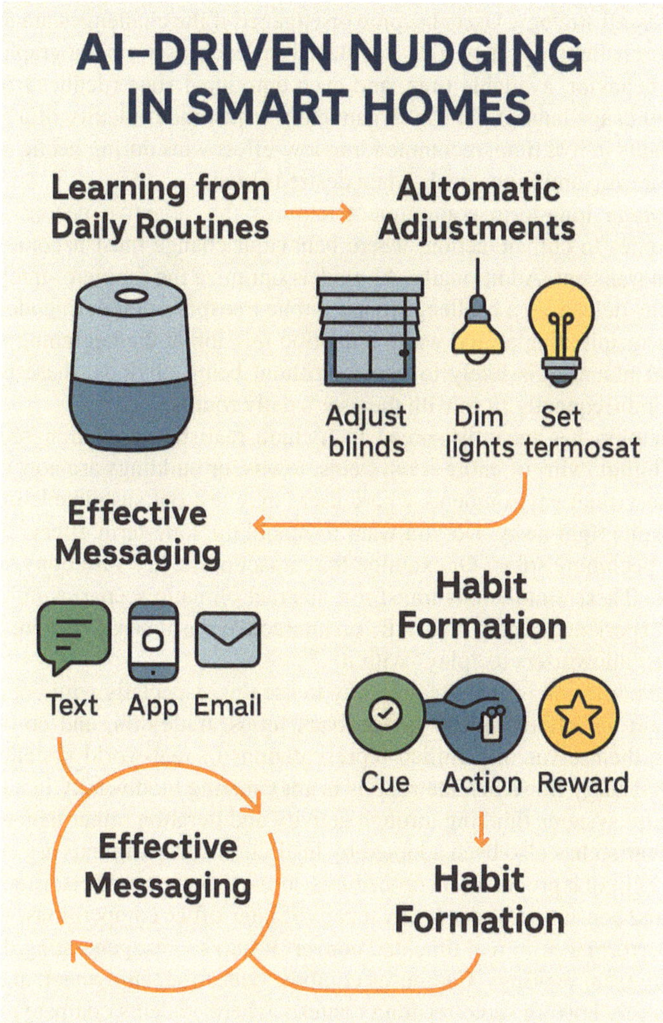

Fig. 18.3 AI-driven nudging in smart homes, where devices like Google Nest learn daily routines to automate adjustments—such as blinds, lights, and thermostats—reducing energy impact. Through personalized messaging and habit loops of cue, action, and reward, AI fosters lasting sustainable behaviours, adapting strategies over time for maximum comfort and environmental benefit

You can compete with coworkers or neighbours, and a leaderboard will show how well you all performed overall. Despite their seeming simplicity, these components can have significant psychological impacts. They provide people a feeling of accomplishment, identification, and community—all of which are essential for forming and maintaining new behaviours.

The power and personalization of gamified sustainability solutions are greatly increased by AI. Conventional games and feedback systems frequently suffer from a one-size-fits-all strategy. Users become disengaged if the challenges are too simple. If it is too challenging, they give up. Based on each user's demographic profile, historical behavior, available time, and even emotional state (deduced from voice tone or app usage habits), AI can dynamically adjust the difficulty of a game. The system adapts in real time, recommending low-effort wins during hectic or stressful times or offering ambitious goals when desire is high.

Maintaining long-term commitment requires this adaptive pacing, especially when it comes to climate action, where behaviour change must become ingrained rather than transient. Additionally, AI models optimize the timing of interaction by, for example, delivering a challenge to use public transportation during ideal weather conditions or initiating a food waste reduction mission at the beginning of a week, when meal planning is likely to occur. Without being obvious, these nudges are gamified and frequently fit in with the user's daily routine.

Gamification has recently spread to include realistic simulation settings. AI-powered digital twins of entire ecosystems, towns, or buildings are now being used to generate virtual environments where users may test out sustainable options and see the results right away. Do you want to model the long-term effects of a neighbourhood becoming solar? Or examine the impact of dietary variations on land use worldwide. These simulations transform abstract climatic scenarios into concrete, relatable experiences using VR, AR, or interactive dashboards. Before the future comes, they allow users to "play" with it.

Climate education is one effective way to use this, especially with young people. Players gain an intuitive grasp of feedback loops, trade-offs, and co-benefits by immersing themselves in gamified representations of real-world systems, such as controlling energy flows in a metropolis or maximizing biodiversity in a landscape. They acquire systems thinking through activity and iteration rather than textbooks.

Gamification has also been adopted by institutions. Departments are encouraged to embrace circular procurement procedures, move to plant-based catering, minimize printing, and consume less energy by means of inter-office competitions. AI systems monitor performance in real time and convert it into scores, generating dashboards that display team rankings. Campuses combine education, enjoyment, and quantifiable impact by hosting semester-long contests where students compete to conserve water, ride bicycles, compost food waste, or attend climate literacy programs. These initiatives promote a sustainable culture that is strengthened by peer accountability and acknowledgment.

In this area, several platforms have become leaders. For example, JouleBug uses an app that tracks behaviours and gives points for eco-friendly behaviour to gamify sustainable habits (Fryers, 2017). Users can create teams, share their progress, and participate in themed challenges thanks to its social features. Eco-feedback games have been included into municipal energy programs in Japan and the Nordics, transforming households into climate actors who use in-game cash to vote on local green initiatives, unlock rewards, and receive weekly performance reports.

Millions of players already enjoy city-building games like SimCity and Cities: Skylines, which are now including sustainability trade-offs and climate scenarios (Jolly & Budke, 2023). In addition to designing heat-resilient urban layouts and responding to floods, pollution, and energy crises, players must strike a balance between economic expansion and renewable infrastructure. These games transform from amusement into civic laboratories that educate the next generation of climate-conscious decision-makers thanks to AI-driven complexity.

What unites these examples is the idea that climate action need not be a grim duty. It becomes something people choose to do—together—when it is reframed through play, creativity, and competition. Gamification demystifies the situation rather than trivializes it. Additionally, AI-powered solutions guarantee that sustainability issues continue to be approachable, flexible, and fulfilling, assisting people and communities in transitioning from passive awareness to joyful, active engagement (Matindike & Matutu, 2025). Games might be one of the most important tools we have in a world that is devoid of optimism and overtaken by complexity.

18.4 Designing for Motivation, not Manipulation: Ethics and Inclusion

A basic query emerges when gamification and AI-powered nudging tools become more integrated into daily sustainability: where does manipulation end and incentive begin? Designing for climate effect entails influencing human behaviour, but doing so morally calls much more than simply incentive structures and persuasive algorithms. It calls for openness, tolerance, and respect for personal autonomy. Gamification must never be used as a cover for coercion.

The ethical boundaries between nudging and manipulation lie in intention, consent, and control. AI systems work in a sphere of influence when they adjust to human behaviour, pick up preferences, and present tailored challenges or suggestions. This is not necessarily bad; individuals like being reminded to conserve money, cut back on waste, or maintain their health. However, there is a difference in whether individuals have control over their decisions, are aware of how the system operates, and can opt out. Ethical design makes ensuring that incentives and feedback are used to further the interests of the user, not just corporate or institutional purposes masquerading as green.

Transparency is essential. Users ought to understand what information is being gathered, how behavioural models are created, and how gamified incentives or nudges are determined (Afshar Jalili, 2020). Trust-building systems are transparent about their goals: to promote rather than to impose. They enable users to establish their own priorities and alter the degree of participation. When climate action is embraced rather than forced, it has the greatest impact.

Consent needs to be continuous and significant. All too frequently, digital systems use ambiguous language or passive checkboxes to get consent. Consent in gamified sustainability platforms should include the option to compete, share statistics with peers, or take part in challenges. It should be free for users to change their objectives, suspend conversations, or stop participating completely. Players that feel empowered are driven players.

Autonomy is crucial, but so is equity. If AI-driven gamification is created without consideration for inclusivity, it can readily widen the digital divide (Cheshmehzangi, 2025). Not everyone has access to smart home appliances, mobile phones, or fast internet. Different communities have different motivators. For example, what motivates a suburban commuter to cycle to work might not be the same for a refugee in a megacity or a rural farmer (Gosnell & Abrams, 2011). A variety of motivational profiles, language diversity, and cultural significance must all be taken into account. This entails creating adaptable, modular systems that provide a variety of access points, incentive schemes, and success stories.

In certain situations, gamification must be community-based rather than individual-based, emphasizing group efforts, regional pride, or intergenerational objectives. In others, it could be necessary to reward participation as well as behaviour modification, creating space for exploration, learning, and failure without passing judgment. By identifying various engagement styles and customizing feedback to promote advancement in a variety of respectable methods, AI can assist these paths.

Ultimately, gamification is valuable for what it makes possible rather than for fun. Behavioural design driven by AI isn't about deceiving people into "doing the right thing". It's about lighting the way and making the next sustainable decision more obvious, attainable, and fulfilling. It transforms intangible climate goals into real, everyday missions. In the face of a massive global crisis, it makes room for happiness, community, and experimentation.

The last realization is straightforward but profound: gamification is about playing forward rather than merely playing games. promoting a culture in which taking action seems feasible. where everyone sees themselves as contributing to the creation of a liveable future, regardless of background or skill. where AI establishes the framework for change rather than controlling behaviour. Even the tiniest acts can have a profound impact on systems of change when they are driven by an ethical, inclusive, and purpose-driven motive. And everyone counts in that game.

References

Adler, S., Hitzig, Z., Jain, S., Brewer, C., Srivastava, V., Christian, B., & Trask, A. (2024). *Personhood credentials: Artificial intelligence and the value of privacy-preserving tools to distinguish who is real online*. Available online from: https://ora.ox.ac.uk/objects/uuid:d5e462dc-0e60-43f7-a583-d4c528d14d45

Afshar, Y. (2020). I rather share my knowledge: Applying gamification approach and nudge theory to develop an incentive system. *VINE Journal of Information and Knowledge Management Systems, 50*(2), 203–217.

Ajayi, T. (2025). *AI-powered meal planning app to support good health*. Available online from https://www.theseus.fi/handle/10024/889521

Alsawaier, R. S. (2018). The effect of gamification on motivation and engagement. *The International Journal of Information and Learning Technology, 35*(1), 56–79.

Cheshmehzangi, A. (2025). Intersecting paths: ICT's role in driving innovation for a sustainable future. In *Urban innovation and sustainability: ICT-based interventions in Hainan, China* (pp. 163–174). Springer Nature Singapore.

Cheshmehzangi, A., & Dawodu, A. (2018). Stakeholder methods to climate change and sustainable development. *Sustainable urban development in the age of climate change: people—The cure or curse* (pp. 23–68). Springer.

Fryers, G. C. (2017). *Gamification for environmental change?* (Master's thesis). Royal Roads University.

Gosnell, H., & Abrams, J. (2011). Amenity migration: Diverse conceptualizations of drivers, socioeconomic dimensions, and emerging challenges. *GeoJournal, 76*(4), 303–322.

Jolly, R., & Budke, A. (2023). Assessing the extent to which players can build sustainable cities in the digital city-builder game "Cities: Skylines." *Sustainability, 15*(14), 10780.

Kim, J. E. (2017). Fostering behaviour change to encourage low-carbon food consumption through community gardens. *International Journal of Urban Sciences, 21*(3), 364–384.

Lehto, T., & Oinas-Kukkonen, H. (2011). Persuasive features in web-based alcohol and smoking interventions: A systematic review of the literature. *Journal of Medical Internet Research, 13*(3), Article e1559.

Matindike, R., & Matutu, D. N. (2025). AI for community engagement and environmental awareness: Harnessing artificial intelligence to promote sustainable practices in the global south. In *AI and ecological change for sustainable development* (pp. 383–414). IGI Global Scientific Publishing.

Mohammadian, B. (2024a). Smart nudge marketing: AI-driven behavioural interventions for sustainable tourism. In *Proceedings of the International Management Conference, 18*(1), 529–539. Faculty of Management, Academy of Economic Studies, Bucharest, Romania.

Mohammadian, B. (2024b). AI-augmented nudges: A new frontier for tailored interventions in sustainability. *Creativity and Innovation in Digital Economy*, 15. University of Tehran.

Chapter 19
Voices in the System with Algorithmic Participation: AI for Inclusive Climate Governance, Civic Engagement, and Policy Co-Creation

19.1 The Democratic Deficit in Climate Policy

One of the most enduring problems in climate governance is not merely technical; it is also democratic, even after decades of climate summits, discussions, and commitments. Too frequently, decisions on climate policy are made in rooms remote from the people who will be most impacted by them. Indiagenous groups, marginalized communities, young voices, and citizens without official authority continue to be underrepresented in planning and disenfranchised from the procedures that will determine their future. Both morally and practically, this democratic deficit occurs when decisions are made about individuals without their meaningful participation. Climate solutions lack resilience, inventiveness, and legitimacy in the absence of inclusive engagement. Instead of being co-created commitments, they become enforced policies.

This lack of civic voice exacerbates the disconnect between public beliefs and policy action. People have strong feelings about climate justice, their local communities, and future generations, yet formal systems frequently ignore or fail to document these feelings. Even when there are channels for participation, including neighbourhood meetings or public consultations, they are typically inaccessible, poorly attended, or at best symbolic. The policy is written in technical terms. The procedures are laborious. The opportunities for influence are limited. The complexity, urgency, and diversity of lived experience in a warming world are difficult for climate governance to adequately represent at this time.

At the same time, the climate emergency necessitates more sensitive, quick, and flexible decision-making than in the past. Instead of planning for months, cities need to react to heatwaves with days' notice. Local demands, worker disruption, and regional equity must all be taken into consideration throughout national energy transitions. When and how they move must be decided by coastal communities. Infrastructure that can listen at scale, react in real time, and incorporate feedback from everyone—not just experts—is necessary in this new era of climate policy.

© The Author(s), under exclusive license to Springer Nature Singapore Pte Ltd. 2025

A. Cheshmehzangi, *AI Agency in Climate Emergency*, Urban Sustainability,
https://doi.org/10.1007/978-981-95-4192-8_19

This is where AI comes in, not to replace democracy but to enhance it. AI has the potential to broaden the scope of who can be heard, how involvement occurs, and how policies change in response when presented as a force for climate urgency. AI can identify the signals present in millions of online posts, community feedback surveys, spoken word transcripts, and civic hackathons thanks to extensive sentiment analysis, natural language processing (NLP), and pattern recognition. The result is a patchwork of dispersed public knowledge, including priorities, ideas, concerns, and fears that were previously dispersed, unorganized, and imperceptible to decision-makers.

AI makes involvement possible even when one is not physically present. While not everyone is able to attend a town hall, anyone can use platforms, applications, or SMS-based systems to share their lived experiences, express their opinions, or offer a local perspective online. Governments and institutions may use AI to assess such information and mine it for direction, urgency, and subtlety rather than just sentiment ratings. We cannot afford to overlook what the public already understands, experiences, and hopes for in a climate emergency. AI provides the tools to continuously and deeply listen.

A living system of data-driven co-creation replaces the static rituals of participation, such as tokenistic feedback forms, yearly hearings, and occasional referenda. Civic AI systems that identify blind spots in adaption planning, identify equity issues in real time, or model community reactions to planned infrastructure can all help improve climate policy. Social emotion and civic foresight can be used to simulate policy situations in addition to environmental data. What if climate authorities could anticipate public reluctance or support for green investments rather than reacting to criticism after the fact? What if planning procedures were open-source algorithms that were taught from community knowledge?

There are growing examples. Cities are allocating funding for climate adaptation using AI-assisted participatory budgeting methods. Municipal authorities can determine which communities feel left behind in green transitions by using sentiment-mapping dashboards. Machine learning-powered youth-led internet platforms are bringing together innovations and needs that are rarely able to find their way through institutional channels. Instead of using bureaucracy, the new infrastructure of participation is being constructed using data, analytics, and inclusive design (see Fig. 19.1).

Integrating algorithmic participation within the framework of climate governance is the obvious challenge at hand. should acknowledge AI as a weapon for democratic renewal during a global crisis, rather than only as a tool for efficiency. In the face of societal disintegration and ecological degradation, inclusion is essential—not optional. People are more inclined to believe in, support, and maintain solutions when they are involved in their development. And we start to create a system of governance that is not just smart, but also just, when AI is utilized to reinforce that voice rather than silence it.

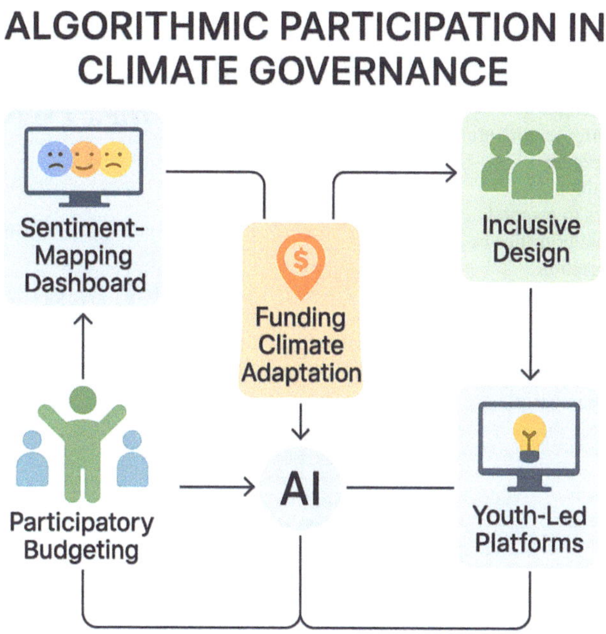

Fig. 19.1 AI-assisted participatory budgeting empowers communities in climate adaptation decisions. Sentiment-mapping dashboards reveal underserved areas, while youth-led platforms connect grassroots innovations with institutional needs. By integrating inclusive design, AI becomes a force for democratic renewal, ensuring climate governance is not only efficient but also equitable, representative, and trusted by the people it serves

19.2 Listening to the Public: AI for Sentiment Analysis and Civic Insight

Opinion, experience, and passion abound in the public arena in this age of societal fragmentation and climate volatility. People use hashtags, blog posts, community forums, and online petitions to express their concerns about floods, their annoyance with growing energy costs, their desire for green jobs, and their mistrust of top-down plans—often without attending formal hearings or submitting policy proposals. If we have the means to listen, there are strong messages to be found in this vast, disorganized, and ever-changing corpus of civic expression. These dispersed voices can be transformed into structured civic insight with the help of AI, especially through developments in Natural Language Processing (NLP). This will give planners and policymakers a more in-the-moment understanding of the social dynamics driving climate action.

Whether it's an open-ended response in a government survey, a petition against an infrastructure project, or a local Facebook group debate on heat stress, natural

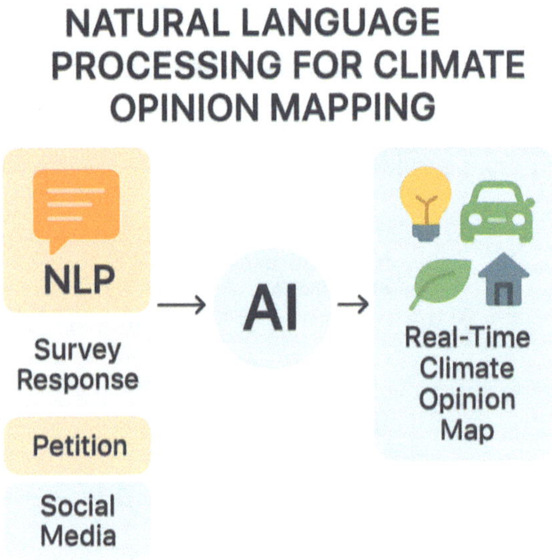

Fig. 19.2 AI-powered natural language processing transforms surveys, petitions, and social media into a real-time climate opinion map. By detecting sentiment, trends, and emerging concerns, it enables decision-makers to respond quickly to public priorities in energy, transportation, adaptation, and environmental justice—strengthening inclusive climate governance and targeted community engagement

language processing (NLP) allows robots to comprehend and interpret human language (Bastianelli et al., 2014). AI can decode not just the content but also the emotion, intensity, and trends that expose public priorities and pain spots by quickly processing enormous volumes of such data (Rathi, 2025). This is revolutionary in the context of climate governance. It enables decision-makers to interact with a dynamic, changing map of social concern and desire rather than limited polls or sporadic focus groups (see Fig. 19.2).

The real-time mapping of public opinion across important sectors, including energy, transportation, adaptation, and environmental justice, is one of the most pressing uses of natural language processing in the context of climate change. AI can identify changes in attitudes about renewable energy projects, the introduction of electric vehicles, or the perceived fairness of climate tax schemes, for instance, by examining thousands of tweets. AI can quickly compile citizen reactions during extreme weather occurrences to monitor effects, emergency requirements, and public confidence in the government. These tools aid in determining where there is support for ambitious climate policies and where more extensive community involvement is required in long-term planning.

Additionally, AI is essential in seeing new issues before they become crises or conflicts (Cheng et al., 2024). AI can identify concerns that are emerging at the grassroots level by analyzing conversation across demographics and geographies, tracking the emergence of new terms or phrases, and keeping an eye on changes in tone or emotional expression. This includes narratives that show mistrust of institutions or disinterest in planning processes, as well as false information about climate science, green technologies, or adaptation tactics. Targeted measures, such as improved communication, more inclusive discourse, or the open correction of lies, are made possible by recognizing these indicators.

Governments and civic platforms are starting to use these capabilities globally. Cities have tested sentiment-driven climate dashboards to direct infrastructure roll-outs and communication campaigns. AI techniques are being employed in Africa, particularly in areas with limited institutional data, to convert citizen-generated reports into insights for adaptation planning (Bhanye, 2025). NLP is used by participatory data mapping platforms in Latin America to interpret community narratives and convert them into indices of climate vulnerability. These illustrations demonstrate how AI may bridge the gap between people and policy by converting civic noise into civic knowledge.

Importantly, AI improves human engagement rather than replaces it. With a deeper grasp of what matters most to various communities, it helps planners better prepare for face-to-face discussions. It makes it possible to adjust climate policies in real time to changing public opinion. It creates feedback loops that allow citizens to observe how their opinions influence choices, enhancing trust and accountability.

However, cautious stewardship is also necessary when using AI for civic insight. Transparency, inclusivity in training data, and avoiding the exclusion of less tech-savvy communities or the reinforcement of prejudices are all requirements for algorithms. When using NLP in both local and global contexts, linguistic diversity, cultural nuances, and context-specific interpretation are crucial.

Nevertheless, the promise is potent. AI offers a means to listen at scale—not just to the noisy but also to the quiet but significant—in a world where climate decisions are urgent and public trust is brittle. AI becomes a democratic tool—a listener, interpreter, and link between planetary governance and its constituents—by transforming the dispersed chorus of civic voices into useful information. It serves as a reminder that knowledge is not limited to expert panels or data centers in the context of the climate emergency. AI is assisting in bringing this wisdom to the table where decisions are made. It exists in communities, discussions, and worries.

19.3 Co-Creating Policy: Participatory AI Platforms and Digital Deliberation

The power of communal imagination, in addition to expert panels and institutional roadmaps, is what will shape climate governance in the future. Given the scope and urgency of the climate catastrophe, societies must transition from consultation to co-creation, in which people actively participate in the development of policies rather than merely responding to them (Wibeck et al., 2022). This change is being made feasible by participatory AI platforms and digital deliberation tools, which allow regular people to participate meaningfully in the intricate and frequently technical field of climate decision-making. AI in this context becomes more than just an analytical tool; it also serves as a translator, mediator, and facilitator between systemic planning and civic awareness.

Platforms that allow citizens to participate in the policy-making process are at the heart of this change. Through interactive tools and simulations, these systems enable people and communities to vote on priorities, explore various possibilities, and influence policy. Participatory AI systems enable dynamic venues where ideas emerge, trade-offs are investigated, and different futures can be evaluated instead of providing a static paper for comment. They act as test beds for democratic experimentation, driven by algorithms but rooted in the goals and lived realities of many populations.

The AI-enabled digital twin is one of the most effective technologies in this field. Digital twins, which are interactive, explorable simulations of neighbourhoods, cities, or entire ecosystems that let citizens to see the effects of policy decisions, were first developed as urban planning tools for simulating infrastructure or energy systems (Cheshmehzangi & Pourroostaei Ardakani, 2024). However, they are now being utilized more as civic tools. Are you curious about the effects of various land-use scenarios on your neighbourhood in a floodplain? How does rerouting traffic affect the quality of the air? How does the scale of solar energy adoption rely on rules or subsidies? People may see and feel the effects of different policy options thanks to digital twins. They bring the abstract to life, and AI gives them flexibility, responsiveness, and engagement.

Deliberative AI agents that manage the intricacy of public input are layered on top of these platforms to facilitate the co-creation process. AI algorithms can highlight, group, and summarize recurrent themes in a consultation paper instead of allowing thousands of citizen comments to remain unread (Weng et al., 2021). They track changing perspectives over time, find group conflicts and synergies, and make sure that no voice is drowned out. By distilling, these agents increase the inclusivity, effectiveness, and transparency of discourse rather than making decisions. They enable people to participate in a meaningful way even if they have little time or informal knowledge (see Fig. 19.3).

Examples from the real world show how these strategies are already influencing climate governance. Through open discussion and voting, AI-powered participatory budgeting platforms in Taiwan have made it possible for residents to allot funds for regional green projects (Cheshmehzangi, 2025; Liu et al., 2023). AI algorithms have been used into digital democracy projects in Barcelona to extract policy recommendations from community forums and manage online debates. In the meanwhile, AI moderation technologies have been employed in the UK and other parts of Europe to manage debate, surface agreement, and synthesize outcomes in real time during climate assemblies, which are randomly selected groups of citizens discussing national climate strategy (Udupa et al., 2023). These instances mark the beginning of a new age of human-centered, algorithmically supported policymaking.

The hybrid model that these platforms facilitate, in addition to the technology, is what gives them their potency. Participatory AI platforms provide a new kind of policy logic by fusing institutional planning models, scientific projections, and grassroots data—such as sentiment, stories, and priorities. One that is emotionally

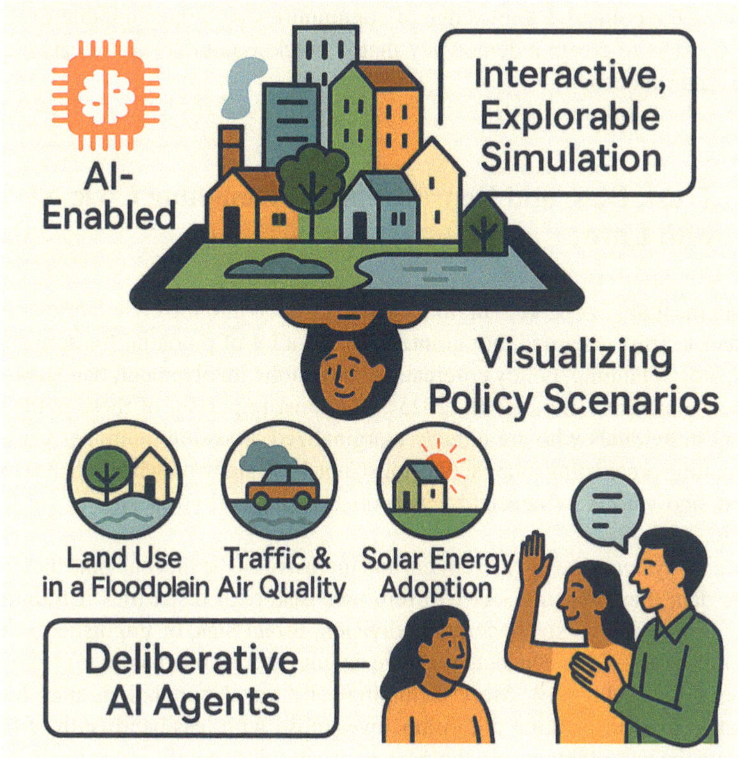

Fig. 19.3 AI-enabled digital twins enhance civic engagement by offering interactive simulations of cities and ecosystems. They visualize policy impacts—such as land use in floodplains, traffic and air quality, and solar adoption—while deliberative AI agents process citizen input, identify themes, track changing perspectives, and promote inclusive, transparent decision-making

sophisticated and rich in facts. One that acknowledges that effective climate policy involves more than just setting emissions goals or cost curves; it also entails matching actions to the needs, values, and support of communities. AI helps governments act sensibly and democratically by bridging this gap.

However, equity, accessibility, and openness must be at the heart of these systems' design in order for them to be successful. Multilingual interfaces, mobile-first accessibility, low-bandwidth functionality, and culturally sensitive design are all need to achieve this (Evanick, 2024). It entails the use of auditable and enhanceable open-source algorithms. And it means integrating human facilitators and community leaders into digital deliberation processes (Cheshmehzangi, 2026); so, AI doesn't replace democratic practice but expands its reach and responsiveness.

Participatory AI-enabled policy co-creation is not a pipe dream; rather, it is becoming a reality. These tools provide a way ahead in a world of climate disruption when credibility and trust are just as important as effectiveness. They assist us in governing alongside us as well as from above. With the people, not just for them. By combining the collective knowledge of communities with the computational power of AI, we start to create a democracy that is quick, equitable, and creative enough for the Anthropocene.

19.4 Trust, Bias, and Representation: Designing Civic AI with Care

One fact must always be kept in mind as AI is woven into the civic fabric of climate governance: trust is earned, not granted. AI has a lot of potential for application in participatory planning, policy co-creation, and public involvement, but it also comes with new concerns (Rizun et al., 2025). The most important of these is the risk of ignoring individuals who are already marginalized, boosting dominant voices, and perpetuating preexisting biases. We must not lose sight of who is heard, who is counted, and who is influenced by systems that they did not design in our haste to innovate.

Civic AI systems are only as ethical and inclusive as the mechanisms they support and the data they consume. AI will reinforce historical disparities if training data reflects them, such as socioeconomic division, urban bias, or linguistic exclusivity (Santorelli et al., 2024). Dialects or native languages may be misinterpreted by sentiment analysis techniques. More technologically vocal populations may be given preference by prioritization algorithms over those with less bandwidth or literacy. Even with the best of intentions, the AI system can subtly codify exclusion by filtering out information that doesn't match clean computational frameworks and converting the loud into the legitimate (see Fig. 19.4).

Fig. 19.4 Civic AI for climate governance, trust must be earned through inclusion and ethics. While AI offers potential for participatory planning and policy co-creation, risks like bias, exclusion, and amplification of dominant voices must be actively addressed to ensure equitable representation and avoid reinforcing historical disparities

Auditability and transparency are essential defences against this danger. Civic AI technologies need to be understandable to the communities they impact as well as to professionals. People need to be able to see how their data is being used, how their participation affects results, and how inputs create outputs. No matter how advanced, black-box models are inappropriate for democratic decision-making. Rather, civic algorithms ought to be subject to public review, development, and critique. This is particularly important in climate governance, as choices affect how people move, eat, work, and even where they reside.

The communal control principle is equally fundamental. The most moral civic AI systems are co-developed with communities rather than just implemented on them (Shukla, 2025). This entails incorporating community-based groups, civic technologists, and local stakeholders in the development, implementation, and advancement of participatory AI platforms. It entails sharing agency in addition to data. It entails considering societal histories, cultural values, and environmental realities when determining not only what AI can do but also what it ought to do.

Every layer of the design must incorporate inclusive elements. Multilingual interfaces that represent linguistic variety and cultural nuances are the first step in this process. It also includes offline engagement tactics that introduce participatory AI tools into community centers, town halls, and schools—settings with a wealth of civic activity but limited access to digital technology. Voice-activated tools for elderly or low-literacy users, visual interfaces for the visually or textually disabled, and interoperable platforms that link mobile apps with radio, SMS, or in-person facilitation are all included. More importantly, participation must be a right rather than a perk of being connected.

Lastly, AI has promise for civic life through amplification rather than automation. AI cannot take the role of discussion, discernment, or debate. It must improve them. It must assist communities in recognizing themselves in the data, bringing to light their common values, and working together to develop climate risk responses that are grounded in lived experience, justice, and dignity. When used properly, AI can assist close the gap between the slow process of democracy and the urgency of the global problem (Duberry, 2022). It can enlarge the room for creativity in public policy, simulate trade-offs before they become errors, and turn fragmented civic opinion into practical information.

But only if we build carefully can this promise come to pass. if we choose digital plurality over technical determinism. If we realize that intelligence is something that is shared by people, places, and stories rather than being the exclusive domain of machines.

Thus, the legitimacy of our policies in the upcoming climate transition will be determined by their creation, by whom, for whom, and with what values—as much as their efficacy. When developed in an open and participatory manner, civic AI provides a new democratic infrastructure that is more equitable, learns more quickly, and listens better. It is the next beginning of engagement, not its conclusion. And this is the future we must create—together—in a warming world where decisions must be fair even if they cannot wait.

References

Bastianelli, E., Castellucci, G., Croce, D., Basili, R., & Nardi, D. (2014). Effective and robust natural language understanding for human-robot interaction. In *ECAI 2014* (pp. 57–62). IOS Press.

Bhanye, J. (2025). Flood-tech frontiers: Smart but just? A systematic review of AI-driven urban flood adaptation and associated governance challenges. *Discover Global Society, 3*(1), 59.

Cheng, Y., Lee, J., & Qiao, J. (2024). Crisis communication in the age of AI: Navigating opportunities, challenges, and future horizons. *Media and crisis communication* (pp. 172–194). Routledge.

Cheshmehzangi, A. (2025). *Urban innovation and sustainability: ICT-based interventions in Hainan, China.* Springer Nature.

Cheshmehzangi, A. (2026). *Generative AI-powered urban digital twins: Pioneering environmental solutions for sustainable intelligent cities.* Springer.

Cheshmehzangi, A., & Pourroostaei Ardakani, S. (2024). Embracing the digital twin paradigm for urban sustainability. In Digital Twin Computing for Urban Intelligence (pp. 1–11). Springer Nature Singapore.

Duberry, J. (2022). Artificial intelligence and democracy: Risks and promises of AI-mediated citizen–government relations. In *Artificial intelligence and democracy.* Edward Elgar Publishing.

Evanick Ed. D, J. (2024, June). Implementing mobile-first strategies in online education. In *The learning ideas conference* (pp. 157–182). Springer Nature Switzerland

Liu, H. K., Guo, Y. W., & Chen, L. Y. (2023, July). Crowdsourcing smart city: SmartTaipei project. In *Proceedings of the 24th Annual International Conference on Digital Government Research* (pp. 44–54).

Rathi, P. (2025). Unveiling the future: An introduction to AI-powered multimedia data processing. In *AI Techniques for Multimedia Data Processing* (pp. 247–268). IGI Global Scientific Publishing.

Rizun, N., Edelmann, N., Janowski, T., & Revina, A. (2025, May). AI-enabled co-creation for evidence-based policymaking: A conceptual model. In *Conference on Digital Government Research* (Vol. 1). https://proceedings.open.tudelft.nl/DGO2025/article/view/991

Santorelli, M., Catullo, D. & Palladino, M. (2024). AI and the reduction of social inequalities in a linguistic perspective. In *Creşterea economică în condiţiile globalizării* (pp. 144–154). Available from https://ibn.idsi.md/vizualizare_articol/218706

Shukla, A. K. (2025, July 7). AI for citizens-empowering public institutions worldwide through inclusive, transparent and ethical AI. *Transparent and Ethical AI.*

Udupa, S., Maronikolakis, A., & Wisiorek, A. (2023). Ethical scaling for content moderation: Extreme speech and the (in) significance of artificial intelligence. *Big Data & Society, 10*(1), 20539517231172424.

Weng, M. H., Wu, S., & Dyer, M. (2021). Ai augmented approach to identify shared ideas from large format public consultation. *Sustainability, 13*(16), 9310.

Wibeck, V., Eliasson, K., & Neset, T. S. (2022). Co-creation research for transformative times: Facilitating foresight capacity in view of global sustainability challenges. *Environmental Science & Policy, 128*, 290–298.

Chapter 20
Catalysts of Change: AI as a Compass in the Climate Emergency

20.1 The Final Threshold: Standing at the Edge of Possibility

The climate emergency is no longer a future scenario; it is our lived present (Allam & Cheshmehzangi, 2024). This is no longer a hypothetical situation for the future. However, a new kind of intellect has been subtly expanding amidst the intensifying heat, storms, and escalating injustices. Dynamic, distributed, and integrated systems of machine learning, sensing, and real-time feedback—not the single, sentient robots of science fiction—can enable mankind to observe, comprehend, and react with a new level of care and accuracy.

This is the real promise of AI in the era of climate change: it will enhance our ability to act quickly, sensibly, and justly, rather than replace human decision-making. The exploration of the preceding nineteen chapters has demonstrated that AI is not a single thing, but rather a collection of platforms, tools, and cognitive extensions that are starting to change the face of climate response. At the core of this change lies a potent notion: AI may serve as a compass in a storm, facilitating not only adaptation but also profound, systemic change when it is in line with global objectives.

We are at the beginning of a new era, one that is characterized by enormous promise and existential peril, rather than just the end of a book. The climate emergency is no longer merely a theoretical concept or distant menace. It has already arrived. Floods come sooner, fires burn hotter, and the poorest people in the globe are hardest hit. We are in the age of consequence; the period of expectation is over. However, a different force—one that is not human in form but is created through code, data, and pattern recognition—has subtly surfaced within this developing turmoil. It is a constellation of learning machines embedded in sensors, satellites, urban networks, financial flows, and agricultural plots rather than a single super intelligence. It is AI, which is becoming increasingly integrated into the very systems that determine the future of our planet rather than remaining confined to theoretical speculation

© The Author(s), under exclusive license to Springer Nature Singapore Pte Ltd. 2025
A. Cheshmehzangi, *AI Agency in Climate Emergency*, Urban Sustainability,
https://doi.org/10.1007/978-981-95-4192-8_20

(Baklaga, 2024). It is AI's agency in dealing with some of our calamities; yet, not alone, and not isolated on its own.

This chapter, and this moment, marks a threshold. We now have tools that can discern complexity with remarkable accuracy: algorithms that map energy flows in real time, identify the signs of fire, drought, or illness, and recognize forest decline before humans do. However, AI is more than just a tool for analysis. It is evolving into a new type of active agent in climate systems. It predicts, models, optimizes, and directs. Furthermore, it is no longer on the verge of action, as we have seen throughout the pages of this book. It is in the process of making decisions. AI is changing how we plan, react, and hope in a variety of ways, from urban planning to climate modelling, from emissions accounting to agricultural precision, and from behavioural nudging to participatory governance (see Fig. 20.1).

Both excitement and prudence are encouraged by this evolution. The use of AI in climate response is a decision, not a given. It can either hasten justice or perpetuate injustice (Mathur & Chamuah, 2025). It may increase asymmetry or democratize data. It can either support extractive pasts or point us in the direction of regenerative futures. How we regulate, integrate, and hold it accountable will determine its course. Now, we need to ask not just what AI is capable of, but also what it ought to do and for whom.

That region has been thoroughly covered in the nineteen chapters that precede this one. They have demonstrated AI's ability to empower decision-makers, map ambiguity, predict risk, and optimize systems (Elayaraja et al., 2024). Perhaps more significantly, though, they have shown a more profound change that is taking place in the way we relate to time, agency, and knowledge. We can no longer depend only on reactive governance or specialized knowledge in the face of rapidly changing global conditions. Systems that can perceive the big picture, adjust in real time, and take action before tipping points are reached are what we need. This is the promise of AI: it will enhance human wisdom rather than replace it.

But let's be clear: wisdom is not the same as precision. Justice is not prediction. Furthermore, data is not neutral. Only when these tools are connected to values, moulded by purpose, and directed by group intent do they acquire significance. Therefore, the last barrier we overcome is not merely technological. It is civic and moral. Is it possible to develop artificial and human intelligences that are sensitive to the intricacy of the biosphere, human dignity, and the need for immediate action?

To successfully pass this threshold, we must accept AI as a warm extension of human ability to care for, repair, and renew rather than as a frigid tool. It is to realize that intelligence is a responsibility as well as a quality. The deepest intellect we can develop in this era of climate emergency is one that recognizes connections, values life, and acts before irreparable harm is done. Thus, may we bring with us a compass of purpose as well as instruments of prediction as we move forward into this uncertain century. Let this last chapter serve as an introduction to the future we still have the courage to create, rather than a conclusion.

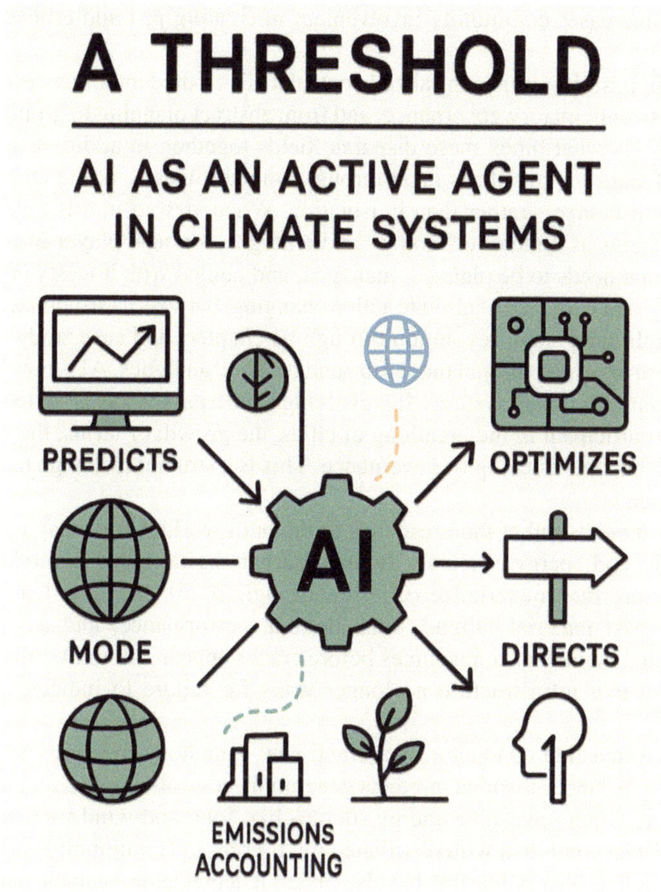

Fig. 20.1 Summary of this critical threshold and AI as the active agent in climate systems, illustrating AI's central role in predicting, optimizing, and directing climate-related actions, linking to domains such as emissions accounting, urban planning, behavioural nudging, and participatory governance

20.2 From Systems to Solidarity: Mapping AI's Emergent Role

We have witnessed AI's transition from the background to the frontlines across industries and regions—from passive data analysis to active engagement in climate-critical systems. AI helps cities and towns keep an eye on their infrastructure before it breaks. It balances renewables in real time in energy networks. It recognizes devastation in forests as it occurs. It provides previously unheard-of precision in directing each drop of water and fertilizer grain on farms. It identifies greenwashing and reorients investments for long-term stability in financial systems. It hears what people have to say and increases community involvement in creating just and efficient climate policy.

A paradigm shift—from centralized control to distributed intelligence, from reactive policy to anticipatory governance, and from abstract planning to grounded, data-rich action—is what binds these disparate fields together, in addition to technical growth. AI's agency manifests itself through embeddedness—in the land, the grid, and public discourse—rather than in isolation. When viewed in this light, AI transcends its status as a mere technology and emerges as a new player in the climate story, one that needs to be planned, managed, and guided with honesty and vision.

Not only is the urgency of climate action changing, but so is its structure. This book has shed light on a silent revolution through its chapters and case study examples: Once limited to computational modelling and backend analytics, AI is now becoming a part of frontline climate systems. It is evolving from a passive observer to a co-actor, a dynamic participant in the breathing of cities, the growth of farms, the movement of markets, and the listening of governance. This is a structural change rather than a symbolic one.

AI now predicts rather than responds in the built world. Cities may redesign for heat, floods, and energy efficiency by using urban digital twins to model climatic stresses before they materialize (Cheshmehzangi, 2026). Machine learning algorithms forecast material fatigue, track thermal performance, and assist in allocating limited maintenance resources before cracks appear—both literally and figuratively—so that infrastructure no longer waits for failure to indicate crisis (see Fig. 20.2).

AI is the invisible conductor of a brand-new symphony in energy systems. AI dynamically balances the load, predicts generation, controls storage, and maximizes grid stability when renewable energy sources like solar and wind vary in time and location. When combined with distributed microgrids and community control, it not only makes clean energy feasible but also makes it dependable, scalable, and socially just.

AI is now the eye that never sleeps in landscapes. Deep learning-powered forest monitoring platforms check for road encroachment, fire risk, and canopy loss with daily precision, frequently identifying infractions before they become more serious. Previously a tedious and expensive procedure, soil carbon estimate now uses spectral and topographic data coupled with AI interpretation to show farmers and policymakers the possibility for regrowth beneath their feet.

Agriculture, one of the most vulnerable and impactful sectors in the climate equation, has been quietly transformed. One plant, one plot, one day at a time, precision farming technologies handle pests, fertilizer, and irrigation at the micro-scale using real-time data and AI decision engines. For farmers who grow the world's food, the outcome is not only increased yields but also more efficient use of inputs, lower emissions, and improved climate resilience.

Finance is also undergoing a reprogramming. AI evaluates the veracity of sustainability claims by searching through financial records, satellite evidence, and business filings (Moodaley & Telukdarie, 2023). It analyzes risk, highlights discrepancies, and assists in allocating funds for adaption initiatives and green innovation. AI serves as

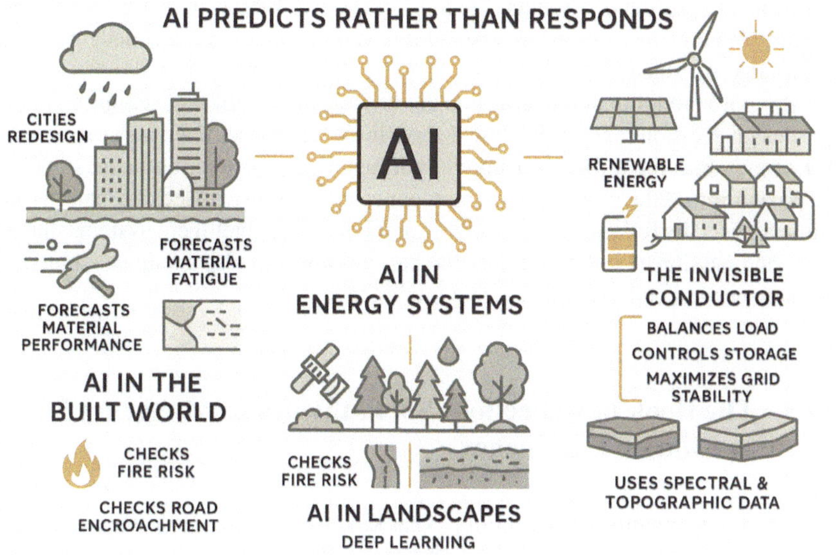

Fig. 20.2 AI no longer waits for the world to change, rather it shapes it. From cities that anticipate heat and floods, to energy systems that balance themselves, to landscapes safeguarded before harm arrives, this is the future we can design. The journey through these pages ends here, but the work begins now: to use foresight, technology, and wisdom to create resilient, just, and thriving worlds for generations to come

a verifier in a world where confidence in green finance is brittle; it is not perfect, but it is incredibly helpful in identifying manipulation and enhancing integrity standards.

Most astonishing of all, AI is starting to function as a civic infrastructure. Natural language processing techniques analyze blogs, petitions, and public comments to uncover patterns of knowledge, concern, and attitude that are frequently overlooked in conventional policy cycles—not to turn people into data points. Communities, from urban neighbourhoods to rural areas, may now participate in climate policy as co-creators rather than merely recipients thanks to sentiment dashboards, participatory platforms, and deliberative tools. In this way, AI acts as a bridge—between people and planners, between technical intricacy and lived reality, and between institutions and individuals.

Not just an increase in processing power or data accessibility is what connects all these changes. The logic of governance is changing from one that is centralized, isolated, and reactive to one that is distributed, anticipatory, and participative. AI is becoming an embedded actor that is positioned within ecosystems, infrastructures, economies, and communities rather than a remote controller.

AI's emergent agency comes from this embeddedness. Its intelligence is contextual rather than generic. Its power is networked, not abstract. The difficulty is becoming more than just technological as this agency expands; it is also moral, political, and profoundly human. We have to choose: Who is in charge of this intelligence? Who gains from its understanding? Whose voices are encoded, and whose knowledge is valued?

AI is not a monolith. It is a reflection of our societal priorities, governance systems, and design decisions as a whole. It will serve limited purposes if we frame it narrowly. Yet, it may become a force for unity—not just across sectors, but across lives—if we design it to reflect the diverse, interconnected, and justice-seeking nature of the climate crisis. Its full promise lies in this: not only more intelligent systems, but also more compassionate ones. Not only the best answers but also group accountability. We now need to cultivate this climate intelligence.

20.3 The Book in Reflection: Six Pathways to AI Agency in Climate Emergency

This book began with a provocation: What kind of intelligence do we need for the climate emergency we now inhabit? The answer has become more apparent across its twenty chapters—not a single type of machine cognition, but a multi-layered, intentional, and collaborative intelligence that addresses the complexity, urgency, and moral significance of climate breakdown.

At its core, this is a book about agency. Not just the agency of human-made (or artificial) systems, but also of ecosystems, communities, and human institutions that are enabled by them. We have seen AI function as an enabler, integrating long-separated efforts, accelerating what is achievable, and uncovering what is concealed, rather than as a substitute for climate action. AI improves our perception, decision-making, and response—faster, more fairly, and more comprehensively—whether in adaptive energy grids, regenerative farming, early warning systems, or real-time deforestation alarms.

Another important realization is that the value of AI is totally dependent on how it is integrated. When intelligence is separated from reality, it becomes inactive. However, it becomes catalytic when placed in urban areas, forests, financial systems, and public forums. It provides leverage rather than just knowledge. Instead of reacting, use leverage to predict. to match climate goals with rewards. to allocate limited resources to the most important areas. to make the margins the focal point of the decision-making process.

The principle of transparency is equally important. The more AI influences decisions on infrastructure, emissions, or social behaviour, the more important it is to make sure that systems are auditable, interpretable, and subject to public scrutiny. Without this, opacity, exclusion, and technocratic imposition—the exact problems that climate AI aims to address—run the risk of reoccurring. Thus, the book makes the case that intelligence needs to be accountable in addition to being strong and adaptive.

Inclusivity threads through every domain explored here. The most effective uses of AI are those that incorporate a variety of knowledge systems, cultural viewpoints, and firsthand experiences, such as AI-informed policy co-creation or participatory urban planning tools. This is not only morally imperative, but also practically required. Climate systems are social, living, and very local. AI will miss the very dynamics it is supposed to comprehend if inclusive design is not used.

The book concludes by demonstrating that the promise of AI is ultimately relational. It has nothing to do with control. Coordination—between individuals, information, systems, and values—is the key. AI becomes more than just a tool when it is in line with intergenerational stewardship, ecological regeneration, and climate justice (Halsband, 2022). It turns into a link between imagined and frightened futures. between what we already know and what we haven't done yet. The most important lessons learned are that intelligence must support life rather than merely simulate it. that rather than being treated like commodities, climate solutions need to be managed as commons. Furthermore, the responsibility with which that agency is employed defines it, not just its capabilities, whether it be human or artificial.

Although we face a huge task, the transformational architecture is clearly apparent. The correct AI—intelligence that is in line with planetary bounds, grounded in fairness, and informed by human values—is what the climate emergency requires, not more AI. It is our responsibility to produce (and somewhat reproduce) this intelligence with foresight that looks beyond the here and now, bravery that faces complexity, and consideration for the delicate, interconnected systems that we are trying to safeguard. By doing this, we regain our agency and pave the way for a future that is worth living for, aiming for, and believing in. We do more than just engineer solutions.

20.4 A Future Worth Building: Intelligence, Justice, and Imagination

So where do we go from here? The conclusion of this book is not a finish line, but a provocation, perhaps a threshold between what is and what could still be. As we have shown throughout the book, the climate emergency is a crisis of systems, governance, equality, and ultimately of imagination rather than just emissions. It is about the collapse of the old and the urgency of the new. AI is neither a panacea nor a techno-saviour in this situation. However, it can serve as a link between disparate facts and a cohesive vision, between actions made in the near term and their long-term effects, between educated citizens and isolated institutions, and between reactive governance and regenerative potential. When applied properly, AI can serve as a glue that ties knowledge to action, community to complexity, and justice to foresight.

To harness AI for climate futures worth living in, one must be unapologetically clear about its guardrails and goals (Debnath et al., 2023; Schütze, 2024). This entails creating mechanisms that are open, auditable, and subject to democratic accountability. It entails incorporating community voices, Indigenous wisdom, and a variety of knowledge systems into the algorithms that are increasingly influencing our shared reality. It entails realizing that togetherness, not just speed, is the solution. Believe ethics and innovation must coexist, and that intelligence should enhance life rather than detract from it. Our AI has to be able to learn context, listen intently, reflect societal ideals, and progress toward justice rather than just forecasting trends or parsing data. Let's create intelligence that respects human dignity and the planet's resilience—not as a post-mortem, but as the foundation of its structure.

Because in the end, this is not a book about technology. It is a book about agency. About the choices we embed in every line of code, the futures we authorize with every dataset, and the courage we summon in this decisive moment. Let's make this our call: o see intelligence not just as computation, but as collaboration. To treat climate not merely as crisis, but as a crucible for collective reinvention. To use AI not as a weapon of control, but as a compass for care. The tools are here, and we seem to be not yet equipped enough for what is ahead. The nexus between the human intelligence, AI, and environmental intelligence could play a major role in how the future may become. The signals are unmistakable. Now is the threshold; the one that we have seen in sci-fi movies before, the ones that we have been eagerly waiting to witness. The most important question of our time is still this: will we rise together, with vision and determination, to form artificial and human intellect as a force of conscience rather than as a convenient tool? Our shared courage is the essential fuel for the future; and more so, guiding this force toward a world that is not only liveable, but honourable, just, and brilliantly alive.

References

Allam, Z., & Cheshmehzangi, A. (2024). *Sustainable futures and green new deals* (pp. 1–137). Palgrave Macmillan.

Baklaga, L. (2024). The role of AI in shaping our future: Super-exponential growth, galactic civilization, and doom. *Journal of Computer Science and Technology Studies, 6*(4), 112–130.

Cheshmehzangi, A. (2026). *Generative AI-powered urban digital twins: Pioneering environmental solutions for sustainable intelligent cities.* Springer.

Debnath, R., Creutzig, F., Sovacool, B. K., & Shuckburgh, E. (2023). Harnessing human and machine intelligence for planetary-level climate action. *npj climate action, 2*(1), 20.

Elayaraja, C., Rahila, J., Velavan, P., Rajest, S. S., Shynu, T., & Rahman, M. M. (2024). Depth sensing in AI on exploring the nuances of decision maps for explainability. In *Optimizing intelligent systems for cross-industry application* (pp. 217–238). IGI Global.

Halsband, A. (2022). Sustainable AI and intergenerational justice. *Sustainability, 14*(7), 3922.

Mathur, V., & Chamuah, A. (2025). Navigating AI and climate change in an unequal world. *Climate Policy*, 1–7. https://doi.org/10.1080/14693062.2025.2503373

Moodaley, W., & Telukdarie, A. (2023). Greenwashing, sustainability reporting, and artificial intelligence: A systematic literature review. *Sustainability, 15*(2), 1481.

Schütze, P. (2024). The impacts of AI futurism: An unfiltered look at AI's true effects on the climate crisis. *Ethics and Information Technology, 26*(2), 23.

MIX
Papier aus verantwortungsvollen Quellen
Paper from responsible sources
FSC® C105338

If you have any concerns about our products,
you can contact us on
ProductSafety@springernature.com

In case Publisher is established outside the EU,
the EU authorized representative is:
Springer Nature Customer Service Center GmbH
Europaplatz 3, 69115 Heidelberg, Germany

Printed by Libri Plureos GmbH
in Hamburg, Germany